АУТИГЕННЫЕ МИНЕРАЛЫ ОСАДОЧНЫХ ПОРОД

AUTIGENNYE MINERALY OSADOCHNYKH POROD

AUTHIGENIC MINERALS IN SEDIMENTARY ROCKS

AUTHIGENIC MINERALS
IN
SEDIMENTARY ROCKS

by

G. I. Teodorovich

Authorized translation from the Russian

CONSULTANTS BUREAU

NEW YORK

1961

The Russian text was
published by the USSR Academy of Sciences Press
in Moscow in 1958.

ISBN-13: 978-1-4684-0654-2   e-ISBN-13: 978-1-4684-0652-8
DOI: 10.1007/978-1-4684-0652-8

CONTENTS

PREFACE

The present work, Authigenic Minerals in Sedimentary Rocks, is designed for the broad circle of lithologists, and also for the geologists and geochemists who are studying sedimentary rocks and ores. Its specific purpose is to stir up interest among lithologists and geologists in the geochemical environment associated with the formation of authigenic minerals in sedimentary rocks, to encourage work in tracing the sequence of formation of these minerals, and to direct attention to other genetic problems.

The book by no means pretends to be a determinative atlas of the authigenic minerals in sedimentary rocks; its task is to draw the reader's attention to questions of origin and, at the same time, to equip him with systematic knowledge about the physical and, especially, the optical properties of these minerals. In addition, the simplified chemical reactions indicated in the book will permit one to distinguish similar minerals, and will also allow him to detect various mineral deposits in the field. Another purpose of the book is to acquaint chemists and geochemists with the properties of the minerals they study in making chemical analyses, minerals that commonly occur as polymineralic aggregates in the samples that are examined.

This work cannot, of course, replace exhaustive treatises on mineralogy in general, such as "Mineralogy" by A. G. Betekhtin [1950] and other texts, or such as books on optical mineralogy (the works of A. N. Winchell and H. Winchell) and others. Its purpose is to give systematic information and to supply auxiliary data, particularly for authigenic minerals of sedimentary rocks, and explain the conditions under which they formed.

At the present time a considerable amount of information has been accumulated concerning authigenic minerals in sedimentary rocks as indicators of the physicochemical conditions at the time of their formation in the sediments or during lithification of the rocks; this matter is discussed briefly in Chapter I. There is an exceptional abundance of new data on sedimentary authigenic minerals obtained recently in the USSR in effecting the five-year plans of developing the domestic economy and through the tremendous expansion of geological prospecting and exploration. This circumstance, as well as the great importance of focusing attention on mineral materials near us, has permitted the author to use reports chiefly of native scientists.

A considerable quantity of material has also been gathered on the problem concerning phases of formation of sedimentary rocks and ores, but there is no agreement on this question; there is a lack of harmony in the concepts concerning the separate stages of rock formation, and it is therefore necessary to generalize the mineralogical descriptions of the successive stages in the growth of sedimentary rocks and ores. The author attempts to answer these questions in Chapters II and III.

Most attention in this book is given to primary chemical minerals and to diagenetic (syngenetic) minerals of sedimentary rocks in the broad sense of the word. Considerable space is devoted to descriptions of the carbonate minerals, glauconite and iron chlorites, sedimentary aluminosilicates of the "clay group," zeolites, iron and manganese oxides and hydroxides, and calcium and iron phosphates.

We are now realizing more and more the important role of colloids, both during transfer of material in the surficial parts of the earth's crust and during sedimentation and mineral development in already-formed rocks. These problems are considered by the author in Chapters IV and V.

If this book helps investigators to acquire, above all, a genetic approach to the study of authigenic minerals in sedimentary rocks and to understand that there is no real genetic mineralogy without geochemistry, the author will consider his efforts justified.

Chapter I

AUTHIGENIC MINERALS AS INDICATORS
OF THE PHYSICOCHEMICAL ENVIRONMENT

The minerals of sedimentary rocks are subdivided into two main groups: authigenic and allogenic. The authigenic components of sedimentary rocks, including authigenic minerals, are the constituents of sedimentary rocks that form at the site where the rocks originate or are found (in the initial sediment or in the consolidated rock). The allogenic components, including allogenic minerals, are those constituents introduced from outside; that is, they formed in a different place and were deposited in the initial sediment in their existing form. Authigenic minerals are of most interest to us since they reflect in great measure the physicochemical conditions of sediment accumulation, the diagenesis of the sediment, and the activity in the lithified rock.

The authigenic character of minerals in sedimentary rocks is recognized by a number of features: 1) idiomorphism — complete crystallographic faces on the grains — or, on the contrary, irregular, intricate grain outlines, which could only have formed in place; 2) irregular crystal forms of hypidiomorphic texture, due to grains growing simultaneously in situ; 3) relations to other allogenic and authigenic components of the sedimentary rock — the formation of rims in place, the phenomenon of recrystallization, or, on the other hand, the replacement of other minerals in the sediment or in the rock, the dissemination of residual inclusions of a mineral (especially an authigenic mineral) at the site where the given authigenic mineral formed; 4) the development of various grains and zones of irregular outline in the cement or the matrix of the sedimentary rock; 5) the lining or filling of pores, cavities, and fractures in the rock; 6) the alternating relationship with other authigenic minerals that form under similar physicochemical conditions, especially in the formation of alternating generations; 7) the features produced by the crystallization of primary colloidal or metacolloidal substances, which may be clearly visible, moderately visible, poorly visible, or very poorly visible (in which case observations should be made with extreme care); 8) the occasional reducing nature of a mineral in underwater deposits, or, on the other hand, a sharp oxidizing character of the mineral in the zone of weathering; 9) the presence of microglobular structure, desiccation cracks in the grains, and other features, indicating that the grains of the investigated mineral were initially colloidal (gel) microconcretions; 10) the characteristic form of crystalline aggregates (radiolitic, axiolitic, etc.); and 11) the replacement of fragmental grains, exhibiting features that exclude the possibility that the replacement occurred during transport.

Sedimentary rocks are divided into four principal groups according to the origin of the authigenic minerals: 1) those of initial chemical origin; 2) the same as (1) but later recrystallized; 3) the products of replacement during diagenesis and epigenesis; and 4) those deposited in open spaces in the rock or in semiconsolidated sediment. A very distinctive group of authigenic minerals forms during the initial stage of metamorphism of sedimentary rocks.

The minerals of sedimentary rocks should arouse our attention as indicators of the physicochemical environment or of a change in the physicochemical environment.

A great number of minerals in sedimentary rocks may serve as indicators of the conditions of sedimentation, of the values of pH, rH, salinity (mineralization and relative content of salt), and, partly, of temperature [Fersman, 1922, 1934, 1937, 1939; Vernadskii, 1925, 1933, 1934, 1936; Pustovalov, 1933, 1940; Sulin, 1935; Teodorovich, 1939_1, 1941, 1946_1, 1947; Sedletskii, $1941_{1,2,3}$; Ginzburg, 1941, 1942, 1946, 1947; Ginzburg and Rukavishnikova, 1951; Krotov, 1943; and Britton, 1936]. In evaluating the significance of authigenic minerals in sedimentary rocks, it is necessary to distinguish syngenetic minerals (more precisely chemical compounds [Pustovalov, 1933]) from epigenetic.

Syngenetic minerals commonly persist during gradually changing conditions (rH, salinity, etc.), continuing to exist in an environment in which they would not form initially.

The rH value, from distinctly oxidizing conditions to distinctly reducing conditions, is best indicated by iron or iron-bearing minerals: iron hydroxides → glauconite → most of the iron chlorites → siderite → pyrite.

In underwater sediments the value of rH generally decreases with depth, a reducing environment becomes more pronounced (and a genetic series of minerals develops); where the rH value is lowest certain mineral indicators will form, but other minerals, mixed as to rH groups, will continue to persist.

There is a well-known series of sediments in basins of varying salinities, beginning with calcite and extending through celestite, $CaSO_4$, and rock salt, to the more soluble salts of magnesium [Kurnakov and Nikolaev, 1938] and of potassium and the double salts. In this series of minerals, corresponding to different concentrations in the water, dolomite commonly occupies a place between calcite and calcium sulfate, as we have shown [Teodorovich, 1946₁], generally overlapping the range of the latter. The formation of $CaSO_4$ modifications depends not only on the salinity but also on the temperature. At 25° $CaSO_4$ precipitates from evaporating sea water in the form of anhydrite if the vapor tension of the solution is less than 17.2 mm; this process occurs during the initial separation of NaCl, in the presence of magnesium and potassium chlorides and sulfates [Van't Hoff and others, 1936]. Gypsum forms when the vapor pressure is less than 14.2 mm and more than 9.2-9.6 mm.

Fluorine-bearing minerals may serve as indicators of standard concentrations in the water. Thus, in the system $CaO-P_2O_5-HF-H_2O$, fluorine is precipitated at first as fluorapatite or as fluorhydroxylapatite, and only "during further concentration of phosphate-free solution is fluorite crystallized" [Kazakov, 1937₁, 1939]. In other words, phosphorites are characteristic of marine waters with normal salinity, fluorite of concentrated marine waters.

By using a series of minerals or individual mineral indicators, we may designate the physicochemical characteristics of the environment during the formation and transformation of sedimentary rocks. Syngenetic minerals may be distributed in a genetic series (see above) or in several series (according to rH, salinity, and other factors); the epigenetic minerals may continue the syngenetic genetic series [Teodorovich, 1946₁]. For example, coarser-grained $CaCO_3$ in calcareous dolomites may be replaced by $CaSO_4$, if the consolidated sediments are buried to a considerable depth, as a result of reaction with highly mineralized waters (formational or connate).

In a normal marine environment, carbonate muds are generally $CaCO_3$, which, in the shelf zone, is commonly replaced by dolomite, partly or completely; the development of dolomite by replacement, apart from other factors, leads to a decrease in the partial pressure of carbon dioxide in the muddy waters in particular and in the atmosphere in general (Precambrian, Paleozoic, and other intervals of time). During the ever-increasing mineralization of waters of marine origin in lagoons, magnesium carbonate is widely developed, dolomite first (especially when the partial pressure of carbon dioxide in the atmosphere is high) and then magnesite; the latter characterizes a further concentration of the lagoonal waters. On the basis of some particular characteristic syngenetic mineral admixtures, carbonate marine and lagoonal deposits, such as the Paleozoic, may be distributed in the following series, reflecting the ever-increasing mineralization of the basin waters: 1) pure calcareous and dolomitic-calcareous deposits of a normal sea, devoid of syngenetic celestite, fluorite, or calcium sulfates; 2) calcareous dolomites and dolomites with syngenetic celestite and fluorite; 3) dolomites with syngenetic anhydrite, celestite, and fluorite; 4) dolomites with syngenetic anhydrite, without fluorite or celestite; 5) dolomites with syngenetic anhydrite and magnesite. Types 4 and 5 may be representatives of primary mottled dolomite-anhydrite rocks (see p. 92). Some dolomites with the mineralogical characteristics of types 4 and 5 may arise by secondary processes, in three stages: 1) the formation of calcareous dolomites in a normal marine environment; 2) subsequent drying and leaching of $CaCO_3$ from the calcareous dolomite during an insular or continental phase; and 3) the penetration of waters from markedly salty lagoons into the sequence of rocks slowly being buried in erosional-reef islands or in small low islands in lagoons.

Of most interest is the problem concerning mineral indicators for pH, most of them being characteristic only of a lower limiting pH value, however, at which value, and above, they begin to form and remain stable [Fersman, 1934, 1937, 1939; Rozhkova and Solov'ev, 1936; Ginzburg, 1942; Teodorovich, 1946₁]. But there are minerals, such as opal, which form and are stable at lower pH values, i.e., beginning with an upper limiting pH value and being stable below. Both the upper and lower limiting pH values have been established for a number of compounds.

For each mineral (chemical compound) there should be distinguished an interval of pH values within which the mineral may form and an interval within which it may exist. With gradual decrease (or increase) in pH the mineral may continue to exist at pH values lower (or higher) than the lower (or higher) possible limit within which it will form. The presence of considerable quantities of several salts in solution may favor the stability of the individual minerals. For example, an abundance of NaCl and magnesian salts in solution increases the stability of dolomite; however, when $MgSO_4$ is abundant and there is a marked predominance of Mg^{+2} over Ca^{+2}, magnesite begins to form. Finally, an important factor is that, depending on the salt content of the water, the limiting pH value at which a given mineral begins to form may shift. The salinity not only influences calcite (carbonates) directly, but is also effective through changes in the equilibrium distribution of the forms of CO_2 (diagram of Buch [1930]). The pH changes differently for various zones of burial within the sediments: normally the pH at the surface part of the sediments is somewhat lower than in the bottom water and in the deeper parts of the sediments themselves [Bruevich, 1938].

With the indicated reservations, we shall note only the principal series of pH mineral indicators:

a) Hydroxides. Iron oxide hydrates precipitate* and are stable at a pH $\geq$ 2.3-3.0; aluminum hydroxide (hydrargillite) forms and persists in sediments in the pH interval from 4.1 to 10.0, though there occurs partial solution of the precipitate at pH values above 7.5-8.0, the process becoming extensive at a pH of 10.0 [Hillebrand and Lendel, 1935]; opal forms in acid, weakly acid, and neutral environments, and it persists into weakly alkaline environments; fibrous crystalline silica and quartz, derived from the recrystallization of opal, persists even in moderate alkaline conditions.

b) Carbonates. Calcite, dolomite, and magnesite are characteristic of alkaline environments **(a pH greater than 7.45), whereas siderite may grow in a neutral medium [Brauns, 1904].

c) Aluminosilicates — "minerals of the clay group." Whereas minerals of the kaolinite group correspond to an acid environment [Noll, 1935, 1936; Sedletskii, 1938], minerals of the hallosite group are more characteristic of neutral and weakly acid environments [Ginzburg, 1941, 1942, 1947; Teodorovich, 1942$_2$; Non-metallic Mineral Resources of the USSR, 1941], and minerals of the montmorillonite and ferrimontmorillonite groups are characteristic of alkaline environments [Noll, 1935, 1936; Sedletskii, 1938, 1939$_1$, 1941$_2$]. In this scheme it is noted [Teodorovich, 1946$_1$] that crystalline minerals of the clay group with a ratio of Al_2O_3 or of R_2O_3 to SiO_2 equal to 1:2 are characteristic of acid environments, those with a ratio of 1:2 to 1:3 of neutral environments (from weakly acid to weakly alkaline), and those with a ratio of 1:3 to 1:6 of alkaline environments. The explanation of this may be found in the increased solubility of silica, especially colloidal, with increase in alkalinity.

During diagenesis of sediments in seas with a normal pH of 8.0 ± 0.5, and in general with a pH of 8.0 ± 1.3 [Kazakov, 1939; Knipovich, 1938; Bruevich, 1938], conditions develop that lead to the predominant growth, and preservation, of minerals in the montmorillonite and hydromica (illite) groups. In soils, according to I. D. Sedletskii, the growth of colloidal-dispersed clay minerals is associated with a given type of soil formation and, above all, with characteristic values of pH [Sedletskii, 1941$_{2,3}$, 1942$_4$]. I. I. Ginzburg [1938] has recognized that weathering zones show, in addition to the effect of climatic factors, a dependence of the growth of clay and other minerals on the chemical-mineral composition of the rocks being weathered.

A mineral may form biogenically, in the skeletons of a number of organisms, in the waters of a basin that may permit it to exist but would not permit it to form by chemical means. Thus, biogenic calcium carbonate in the skeletons of several benthonic organisms may form when the organism lives in water with a pH value below the limit for chemical precipitation; and opal may form when the surrounding medium has a higher pH value than its characteristic range.

In citing the various extreme values of pH, we think it is necessary to state our reservations that the values require further refinement and concrete verification, especially for marine and fresh waters, and also for brines,

---

*In this and all the subsequent examples we assume a sufficient concentration of the corresponding compound in solution, determined by the value of its solubility.

** At first glance this statement may appear too general, especially for magnesite. But it is now known that magnesite may develop and continue to exist in various environments having different pH values.

since there is every reason to suppose that in marine waters, and even more in brines, some of the limiting pH values may be considerably shifted relative to the comparable ranges in fresh water. A characteristic example of this relation is shown by magnesite. In the weathering zone this mineral forms and continues to exist in a markedly alkaline environment, at a pH greater than 9.5 [I. I. Ginzburg], being associated with a saturation of $MgCO_3$ in bicarbonate waters of surficial origin. On the other hand, in an environment of highly mineralized sulfate-chloride briny waters in the Kara-Bogaz-Gol, magnesite is found everywhere with $CaCO_3$ in the bottom muds, where the pH is 7.6 in the bottom waters, as was noted by N. M. Strakhov in 1941 [Strakhov, 1945].

The presence or absence of any particular mineral may also depend on the temperature. Water temperature is reflected, above all, in the precipitation of $CaCO_3$. We have obtained interesting data concerning an intermediate layer of water in the Polar basin that is supersaturated with $CaCO_3$ and that has a higher temperature and higher salinity than normal. However, for $CaCO_3$ to accumulate, the decisive factors, as for many other compounds, are the physicochemical conditions (temperature and others) of the waters at the bottom and in the muds. An abundance of reworked glauconite indicates a relatively low temperature in the bottom waters of the sea and that the waters move slowly.

The geochemical facies determining the conditions of sedimentation and diagenesis in the sediments are of most interest to us. However, during diagenesis extensive and even fundamental changes may occur in the physicochemical conditions, changes that must be detected petrographically or geochemically by the discovery of joint occurrences of minerals or by the presence of features that could only have arisen at different times. Thus, many of the normal marine Paleozoic limestones in the Ural-Volga region contain $CaSO_4$ at depth [Vernadskii, 1933, 1934, 1936; Sulin, 1935] that was deposited epigenetically from metamorphic* and concentrated waters [Teodorovich, 1946_1].

The processes of sedimentary epigenesis are much more significant than many investigators believe. In carbonate rocks epigenetic processes frequently lead to the formation of dolomitic flour or cavernous-porous dolomites by the leaching of $CaCO_3$ from calcareous dolomites and dolomitic limestones [Nikitin, 1890; Teodorovich, 1931, 1946_2, 1950; Solov'ev, 1941; Afanas'ev, 1948], to the dolomitization of limestones by the circulation of ground waters through them [Bogdanova, 1940], or to the dedolomitization of dolomites and the formation of clearly crystalline, so-called secondary limestones [Noinskii, 1913; Krotov, 1925; Miropol'skii, 1930; Teodorovich, 1932; Tatarskii, 1953].

In sandy rocks epigenetic processes especially influence the composition of the cement, which is shown to consist of several generations of authigenic minerals, exhibiting a sequence in their development. These sequential generations of authigenic minerals are recognized by the mutual relations of the grains to each other: corrosion, replacement, and similar features. They indicate a gradual shift in the physicochemical environment during the formation of sandy rocks [Rengarten, 1940, 1950; Ermolova, 1952; Abramova, 1954; and many others].

In some sedimentary ore deposits repeated changes in the epigenetic conditions might have favored even higher concentrations of the ore substance. It is probable that for some deposits, some iron ores for example, the final stage of ore formation was due not only to accumulation and diagenesis of the sediment, but also to epigenesis, especially subsequent weathering (supergene processes).

I. A. Preobrazhenskii [1941] has proposed that "the question of authigenic origin may be raised for every mineral in sedimentary rocks." In this statement he based his view on the growth in place of fragmental grains of various compositions (feldspars, tourmaline, garnet, pyroxene, amphibole, rutile, staurolite, chlorite, etc.), which is generally recognized by lighter borders with idiomorphic outlines.

V. P. Baturin [1942], premising his views on the works of A. E. Fersman, believes that "the most stable minerals in the sedimentary shell are formed from minerals of the deep geosphere in the weathering zone." He includes among undoubted authigenic minerals in sedimentary rocks only quartz, rutile, zircon, tourmaline, and a number of varieties of feldspar (orthoclase, microcline, anorthoclase, and albite). References to other authigenic minerals in sedimentary rocks are believed by V. P. Baturin to be either inexact, such as garnet and other minerals that formed in place during regional or other well-defined metamorphism of sedimentary rocks, or incorrect (hornblende and basic plagioclase), being based only on idiomorphic form of crystals that were, as a matter of fact, of pyroclastic origin. According to V. P. Baturin, the zone of garnet growth apparently extends only to the upper horizons of the metamorphic shell and does not include the sedimentary shell. Baturin has noted the

---

*Tr. Note: Metamorphic waters are ground waters transformed by reaction with the surrounding rocks.

4

presence of the quadrivalent elements Si, Ti, and Zr in authigenic minerals, in quantities greater than their average (atomic) content in the earth's crust; authigenic sedimentary quartz is most widespread, rutile and other forms of TiO_2 next most abundant, and tourmaline and zircon being encountered very rarely. Only the most stable feldspars (potassium feldspar and albite) are found as authigenic minerals in sedimentary rocks.

Investigations of minerals in sandy-silty and argillaceous sedimentary rocks of the epigenetic zone and of previous metamorphism are directly related to this question; such investigations were carried on by A. G. Kossovskaya and V. D. Shutov [1955] in the western Verkhoyansk region and by N. V. Logvinenko [1956] in the Donbas. These studies have shown the importance of epigenetic processes in the transformation of the mineral composition and in the "life history" of sedimentary rocks.

A. G. Kossovskaya and V. D. Shutov studied a section of sedimentary strata in the zone of the western Verkhoyansk region and the adjoining marginal depression, embracing rocks from Lower Permian to Lower Cretaceous and having a thickness in excess of 11,000 m. Four zones were distinguished in this section, each characterized by a different type of cement: 1) a zone of clay cement at the top (upper part of the Lower Cretaceous); 2) a zone of chlorite and chlorite-silica cement next below (lower part of the Lower Cretaceous and all the Upper Jurassic); 3) a zone of quartz-enlargement cement or of quartzitic sandstones (Middle Jurassic-Upper Permian); and 4) a zone of quartz-enlargement and micaceous cement or a zone of phyllitic schists (Lower Permian). The fourth or lowest zone of the section clearly belongs to the zone of metamorphism, regional in this instance. We shall describe the designated zones by the authigenic minerals and by the alterations in the fragmental components.

The upper zone, the zone of clay cement, is characterized by calcium zeolite (laumontite) in addition to clay in the cement, by the hydration of fragmental biotite, and by fragmental ilmenite.

The second zone, the zone of chlorite-silica cement, is characterized by 1) chlorite and opal in the cement (upper part of the zone) or by chlorite and chalcedony or chlorite and quartz (the latter developed in the lower, thicker part of the zone); in addition, laumontite is also present in the upper part of the zone and hydromica in the lower part; 2) the hydration of biotite, occurring only in the upper part of the zone, giving way to the development of chloritized biotite fragments and of "amorphous" biotite, and the development of authigenic chlorite and hydrobiotite; and 3) the presence of fragmental ilmenite only in the upper part of the zone but of fragmental and authigenic leucoxene throughout almost the entire zone (except the uppermost part), and, finally, the appearance of authigenic anatase, brookite, and ilmenite in the lower part of the zone.

The third zone, the zone of quartz-enlargement cement, is characterized by 1) chloride-quartz cement in the upper part, quartz overgrowths throughout the entire zone, and hydromica locally; 2) overgrowths on fragmental quartz and the presence of enlarged (fragmental) and authigenic acid plagioclase, recrystallization of siliceous fragments, the presence of chloritized and amorphous biotite, changing in the lower parts of the zone to cryptocrystalline aggregates of authigenic biotite-like mineral, the development of authigenic chlorite and hydrobiotite (upper part of the zone), and, commonly (throughout the entire zone) the development of authigenic muscovite and chlorite; and 3 the presence of fragmental and authigenic leucoxene, the presence of authigenic anatase, brookite, ilmenite, and sphene in the upper part of the zone, rutile in the lower part.

The fourth and lowest zone, the zone of quartz-enlargement and micaceous cement, is characterized by 1) quartz-enlargement and muscovite-chlorite cement; 2) overgrowths on fragmental quartz and the presence of recrystallized (fragmental) and authigenic acid plagioclase, the recrystallization of siliceous fragments, the development of authigenic muscovite and chlorite or of authigenic muscovite alone; and 3) the presence of authigenic rutile, and also epidote and zoisite, in sandstones in the upper part of the zone.

Zones 1 and 2 should be referred to the sphere of epigenetic development; zone 3 is transitional, exhibiting incipient metamorphism; and zone 4 belongs clearly to regional metamorphism.

Because deposits have been studied in different tectonic environments and because they were not initially completely homogeneous or identical, it becomes difficult to distinguish reliably the zone of epigenesis from the zone of regional metamorphism. The presence of authigenic epidote throughout the entire Lower Cretaceous is a matter that is not clear.

A. G. Kossovskaya and V. D. Shutov believe that the distribution of authigenic minerals, especially accessory minerals (epidote and titanium-bearing compounds), is fundamentally influenced by the initial composition of the fragmental components.

N. V. Logvinenko has recognized an increase in the degree of alteration in the clastic and argillaceous rocks with change in degree of metamorphism of the coal in the Middle Carboniferous deposits of the Donbas and the Dnepr-Donets Basin (i.e., the region of the Greater Donets Basin). He distinguishes three epigenetic zones: 1) an upper or normal epigenetic zone, corresponding to the region of development of long-flame, gas, and brown coals, has clay-carbonate and carbonate cement (in the Dnepr-Donets Basin and the adjoining depression, sands were also noted); 2) a middle zone, or zone of progressive epigenesis, corresponding to the region of development of coking coals, has clay, clay-carbonate, quartz-carbonate-clay, and chlorite cement; and 3) a lower zone, or zone of incipient metamorphism, corresponding to the region of development of lean coals and of anthracite, contains sericite, sericite-quartz, sericite-quartz-carbonate, carbonate, and sericite-quartz-chlorite cement, commonly showing overgrowths.

The upper zone, or zone of normal epigenesis is characterized by a porosity of the rocks ranging from 10 to 20% in the Donbas and up to 25% in the Dnepr-Donets Basin; by swelling of the rocks in water; by only slight development of overgrowths on fragmental grains and only weak formation of kaolinite on micas, hydromicas, and feldspars; by the low degree of chloritization of dark micas and the slight sericitization of plagioclase; by the meager development of authigenic anatase, brookite, and rutile; and by the slight conversion of clay material in the cement of the sandstone to sericite (partly to chlorite).

The middle zone, or zone of progressive epigenesis, is distinguished by porosities from 1.5-2.0 to 8-10%; by the presence of rock that does not generally swell in water; by overgrowths on fragmental grains; by the diminution of the process of kaolin development on micas, hydromicas, and feldspars and of sericitization of plagioclase, by an increase in authigenic anatase, brookite, and rutile; by a low degree of conversion of clay material to sericite (partly to chlorite) in the matrix of pelitic rocks and a medium degree in the cement of sandstones.

In the lower zone, or zone of incipient metamorphism, the porosity of the rocks ranges from parts of a percent to 3-4%, none of the rocks swell in water, overgrowths occur on fragmental grains, and the following processes are pronounced: chloritization of dark mica, sericitization of plagioclase, conversion of the clay minerals in the cement of the sandstones to sericite (partly to chlorite), and the formation of authigenic anatase, brookite, and rutile. It has also been found that the clay minerals in the matrix of the pelitic rocks have been altered to sericite (partly to chlorite) and that authigenic tourmaline is present.

The Middle Carboniferous rocks of the Donbas show deviations from the above-indicated regional scheme of distribution of epigenetic intensities. These deviations are due to the fact that, apart from regional burial of the Middle Carboniferous rocks and the beginning of regional metamorphism, the same rocks in the Donbas were subjected also to the irregular effects of dynamic metamorphism (and, locally, intrusions might be found in the lower strata), and to the effects of hydrothermal activity.

The zones of epigenesis (normal and progressive) as distinguished by N. V. Logvinenko correspond to the upper two zones of epigenesis of A. G. Kossovskaya and V. D. Shutov, whereas the zone of incipient metamorphism of Logvinenko corresponds to the third epigenetic zone of Kossovskaya and Shutov. The lower zone of the latter authors (the zone of phyllitic schists), from the viewpoint of the appearance of regional metamorphism, is situated below the zone of incipient metamorphism of N. V. Logvinenko.

In a general paper in 1955, L. V. Pustovalov, from the data of a number of authors, described new discoveries of authigenic (secondary) magnetite, spinel, rutile, brookite, anatase, feldspars, zeolites and analcime, tourmaline, garnets, staurolite, zoisite, clinozoisite, epidote, chlorite, micas, sphene, sepiolite, palygorskite, and other minerals in nonmetamorphosed sedimentary rocks. He believes that the listed minerals, previously considered to be high-temperature minerals, may form in sedimentary rocks with no accompanying magmatic processes or any clear effects of metamorphism; that is, they may form at the ordinary temperatures and pressures that occur in the sedimentary shell. L. V. Pustovalov emphasized the wide distribution and the important role of sericitization, chloritization, zeolitization, and the formation of analcime during epigenesis in nonmetamorphosed sedimentary rocks; he also noted the extensive discovery of authigenic titanium minerals, feldspar, and palygorskite in sedimentary rocks.

Allogenic minerals are less sensitive than authigenic minerals to changes in the physicochemical environment, especially during lithification and epigenesis of deposits in epicontinental basins. This is due, on one hand, to the great stability of the bulk of the fragmental particles, consisting of quartz and fragments of quartzite, siliceous rock, and quartz silicites,* including quartz schists with mica and mica-quartz schists, and, on the other, to the destruction of the less stable fragmental minerals during transport and during the very process of sediment accumulation (sedimentation). However, during rapid burial of various fragmental material, especially pyroclastic, in geosynclinal regions and intermontane depressions, many geochemically unstable components may appear among the allogenic minerals in sedimentary or volcanic-sedimentary rocks. Subsequently, a considerable part of the unstable fragments will be replaced by other, authigenic minerals, forming both during diagenesis and during epigenesis, i.e., in the lithified rock. The data cited above (pp. 5-6) on studies of the epigenetic zone in sedimentary strata are examples of the latter kind, recording changes in fragmental components (biotite, hydromica, and fine-grained material in the cement or in the matrix of pelitic rocks).

There are many allogenic minerals in sedimentary rocks, especially from the "clay group," that may serve to a certain degree as indicators of pH in the environment, and, at times, may indicate other conditions [Teodorovich, 1939_1, 1942_2; Sedletskii, 1942_2]. For these, one may speak of stability limits of minerals in the surface zone of the earth's crust. As is well known, the range through which a mineral may exist is greater than the range through which it may form.

Fine, clastic kaolinite, hydromica, and montmorillonite, where their alteration is completely absent, may indicate, to a certain extent, the pH value of the muddy waters. An interesting example is found in the distribution of so-called nontronite and serpentine minerals in the Khalilovo iron-ore deposits. We discovered that these minerals, mechanically derived from an ancient weathering zone, generally persist only in the upper horizons of the Khalilovo iron ores, a fact we associate with the high pH values that existed at the time the minerals were formed [Teodorovich, 1939_1]. The pH values, in turn, depended on how near the underlying, partly carbonatized serpentinites were to the surface, the serpentinites representing the basement on which the lower horizons of the Khalilovo iron-ore beds lie. The correctness of this explanation is confirmed by the fact that fragmental chrome spinel, being relatively more stable, is found in all horizons of the Khalilovo ore strata. The presence of serpentine fragments in these rocks indicates the nearness of the respective parental rocks at the time the ore deposits began to form.

---

*A term for siliceous rocks that reflects only their composition independent of their origin; it is used chiefly for quartz rocks whose origin has not been previously determined.

Chapter II

PRINCIPAL PHASES AND STAGES
IN THE FORMATION OF SEDIMENTARY DEPOSITS

It is necessary to differentiate the principal phases and the smaller units, or stages, in the development of sedimentary deposits. We shall begin the consideration of the problem with a survey of the views on the basic phases and stages of development of sedimentary formations prior to incipient metamorphism.

A. Syngenesis and Epigenesis. Sedimentation, Diagenesis, Katagenesis, and Hypergenesis.* Incipient Metamorphism

In the literature on lithology the term "syngenesis" has been firmly established to designate the combination of processes involving accumulation and diagenesis of sediments; the term "epigenesis" is generally understood to embrace the processes of change originating in lithified sedimentary rocks, excluding transformations leading to metamorphism. Despite the somewhat conventional and broad concepts involved in the usage of the terms "syngenesis" and " epigenesis," their usage in the senses indicated should be preserved; they are of practical convenience, since they mark the boundaries of the processes associated with sediments and their alteration, separating these processes from those related to alteration of sedimentary rocks in the surface zones of the earth's crust (at relatively high temperatures and pressures).

The term " syngenesis" was proposed in 1922 by A. E. Fersman, who, speaking more precisely, distinguished synchronous and diagenetic aqueous sediments, i.e., syngenetic and diagenetic minerals (in sediments). Syngenetic and diagenetic processes together embrace the formation of sediments on the floors of aqueous basins and the conversion of these sediments into solid rocks. However, A. E. Fersman definitely differentiated these two periods in the development of sedimentary rocks: the deposition of the sediments themselves, and the diagenesis of the sediments. "The minerals that formed as constituent parts of the sedimentary process are called primary (synchronous). For example, $CaCO_3$ in shells or beach oolites, and pisolites in bog iron ore in lakes, as well as other such forms, are of this type. However, the second epoch in the developmental history of the rock, which we term diagenesis, is much more important to us. This term designates a combination of all those processes originating on the floor of aqueous basins in the primary muddy or still unconsolidated sediments, processes that convert the sediment to rock directly under the surface of water before a new layer of sediment is deposited" [Fersman, 1922, p. 29].

It is now considered an established fact that diagenesis (diagenesis of sediments) generally continues for a much longer time than that involved in the deposition of a single layer (that is, that passes before the deposition of a new layer); such rapid consummation of the diagenetic processes has been found only in certain calcareous sediments that immediately form a solid mass on the floor of the basin (a reef framework, the biogenic structure of which forms a calcareous incrustation on the floor of a marine basin in the region of strong bottom currents).

L. V. Pustovalov, in his well-known paper of 1933 on geochemical facies, pointed out that the following may be distinguished in each sedimentary rock: 1) a fragmental or clastic part; 2) syngenetic minerals of organic and chemical origin, associated with the developmental stage of the sedimentary rock, arising (precipitating) primarily from solution during accumulation of the sediment and during diagenesis of the sediment; and 3) secondary or epigenetic formations, arising in the lithified rock. This broad concept of syngenesis is confirmed in the literature on lithology.

---

*Translator's note: American geologists use the word "supergene" instead of "hypergene." However, this paper refers to Fersman's usage of the term and to the significance of the term, and, because of this, it is thought better to retain the Russian form. It hardly seems proper to say "Fersman conceived supergene to include . . ." when he actually said "hypergene".

The term "epigenetic" has been used in the petrography of igneous rocks to designate secondary alteration in minerals; epigenetic mineral deposits have also long been distinguished, the deposits having developed at a time different from the host rock, i.e., forming after the host rock by replacing it or by filling fractures and cavities.

Thus, the minerals listed by L. V. Pustovalov as growing in lithified sedimentary rocks within the earth's sedimentary shell are properly called epigenetic.

At present, epigenesis in lithology is conceived to designate the processes at work in already lithified sedimentary rocks in the upper shell of the earth's crust, i.e., the stratisphere. This so-called clearly secondary process (or phenomenon) is sometimes called late diagenesis, or diagenesis of rock. Some authors have recently proposed other terms in place of "epigenesis" to designate the processes of transformation of already lithified sedimentary rocks: "metadiagenesis" (M. S. Shvetsov in 1956-1957) and "metagenesis" (N. M. Strakhov in 1957). It seems to us superfluous to introduce these terms. Furthermore, the terms "metagenesis" and "metagenetic" are synonyms for "epigenesis" and "epigenetic," as is well known [Loewinson-Lessing and Struve, 1937].

The term "diagenesis" is considerably less definite. It is derived from the Greek work "diagenes," which means regeneration." It was first used by Gümbel to designate all the changes in sedimentary rocks (both during accumulation of the sediments and during the transformation of these sediments — metamorphism) in the course of their conversion to crystalline schists. Later the term "diagenesis" came to be used [Walter, 1893-1894; and others] to designate only processes involved in converting sediments into rock, and, by some authors, to refer to all processes arising during lithification and transformation of sedimentary rocks, except for processes that are clearly metamorphic. Finally, M. A. Usov [1924] extended the concept of "diagenesis" to igneous rocks.

According to A. E. Fersman, diagenesis ceases when the sediment on the basin floor becomes covered by a layer of different composition; diagenesis then gives way to katagenesis. Fersman [1922] conceived katagenesis to be the combined transformations of sedimentary rocks that occur after the sediment has been isolated from the aqueous basin by a layer of new sediment, until the sediment becomes land surface and is exposed to the atmosphere.

Twenhofel [1936] has considered diagenesis to be the changes in sediments occurring prior to its lithification; he excludes all changes in consolidated sedimentary rock that begin after uplift, that begin simultaneously with the initial action of subsurface waters of meteoric origin.

In 1934 M. S. Shvetsov noted the lack of precision in the term "diagenesis," the desirability of rejecting it, and the absence of any term to take its place; he consequently proposed to distinguish "primary diagenesis" or "diagenesis of sediments" from "late (posthumous) diagenesis" or "diagenesis of rocks." Shvetsov [1934] conceived diagenesis of rocks to signify changes in rocks that had been raised above the level of the basin in which they formed; he excluded the processes of surface weathering from late diagenesis. In 1948 M. S. Shvetsov wrote [p. 36]: " We shall consider all sediments to have become rocks when they have been shifted to a position markedly different from that in which they were deposited and when they are no longer subjected to the processes affecting them at the site of their deposition."

L. V. Pustovalov [1940] named the stage of conversion of sediment into rock the "stage of early diagenesis," or the diagenesis of sediments; he also called it the "stage of syngenesis." Like M. S. Shvetsov, L. V. Pustovalov believes the term "diagenesis" to have become in great measure weakened because of the various views concerning it that are held by different authors. Considering the wide use of the term "diagenesis," L. V. Pustovalov employed it in the broadest sense, i.e., in application to all rocks, but, in relation to sedimentary rocks, he distinguishes between "early diagenesis" and "late diagenesis," as M. S. Shvetsov does.

At present most lithologists generally use the term diagenesis (diagenesis of sediments) to indicate the combination of all processes (chemical, physical, physicochemical, biochemical, and geologic) that have controlled the conversion of sediments into consolidated rock (generally) without the accompaniment of orogenic forces or internal heat of the earth. It includes compaction and desiccation of sediments, cementation, recrystallization in part, leaching of salts, and related processes. Diagenesis of sediments is sometimes called early diagenesis.

Katagenesis is considered by many lithologists to include the complex of chemical, mineralogical, and physical transformations in lithified sedimentary rock in response to subsurface waters.

Hypergenesis (supergene activity) was used by A. E. Fersman [1922] to indicate processes of surface alteration (weathering) of rocks, i.e., the combination of chemical and mineral-forming processes occurring in the surface segments of the earth's crust, at the boundary between the lithosphere and the atmosphere or at the boundary between the lithosphere and the hydrosphere. Ancient hypergenesis refers to changes in minerals and rocks that occurred during breaks in sedimentation. At present hypergenesis generally is applied to phenomena of surface weathering that occur on land.

In using the general terms "syngenesis" and "epigenesis" to designate the principal phases of lithification of sedimentary deposits, we believe it necessary to make even more detailed subdivisions in the processes of developing sedimentary rocks, distinguishing smaller steps, or stages, corresponding approximately to the terms of A. E. Fersman (except for the boundary between diagenesis and katagenesis). The necessity of delimiting stages in the formation of sediments and stages of diagenesis is quite properly emphasized in the papers of L. B. Rukhin [1953] and N. M. Strakhov [1953].

L. B. Rukhin distinguished three stages of lithification (more precisely, three stages in the developmental history of sedimentary formations): syngenesis, diagenesis, and epigenesis.

According to L. B. Rukhin, syngenesis "combines the processes originating in the uppermost part of the sediments during the first phase of their existence. These processes, in substance, occur even in the middle of the deposits" [1953, p. 203]. Diagenesis, according to L. B. Rukhin, "is characterized by processes arising in the sediments themselves and lead to their conversion into rock" [idem]. In Rukhin's discussion, syngenesis may include phenomena of submarine weathering (halmyrolysis), which are not everywhere developed; he points out that syngenesis in normal shales is delimited by a thin surface layer of sediment.

N. M. Strakhov wrote [1953, p. 12] "As is well known, three fundamental stages may be distinguished in the history of a sedimentary rock: a) sedimentation, or the stage of sediment deposition; b) diagenesis, or the stage of conversion of sediment to consolidated rock; and c) epigenesis, or the stage of alteration of already consolidated rock, but not including metamorphism or weathering."

It is necessary here to consider in more detail the concept of sedimentation or syngenesis according to A. E. Fersman. Apparently two basic types of subaqueous sedimentation should be distinguished: a) one in which the sediments chiefly fall immediately to the floor of the basin and afterward remain almost undisturbed; and 2) one in which particles that have already fallen to the floor are repeatedly stirred up during the stage of sediment accumulation.

The first type may be called passive, or undisturbed, sedimentation; the second, active, or continuously disturbed, sedimentation; this second type is apparently considered to be chiefly "syngenetic" by L. B. Rukhin.

Finally, M. S. Shvetsov recently [1957] called the attention of lithologists to processes that originate in the surficial zone of sediments, generally of the subaerial type, where the sediments do not pass through the normal stages of diagenesis of subaqueous type. He proposed that these processes be designated by the term exogenesis. He cited loess and loess-like rocks in general as examples of sediments that had passed through exogenesis and had been converted to independent rocks.

The discrimination between a stage of sedimentation, corresponding to syngenesis of A. E. Fersman, and a stage of diagenesis is, of course, proper. However, for the present purpose of delineating the complete history of sedimentary formation, it is necessary to go even farther in the indicated direction and to distinguish subsequent phases and stages in the existence of the rocks and in the corresponding groups of authigenic minerals.

L. B. Rukhin [1953] proposed that a distinction be made between progressive and regressive epigenesis, and M. S. Shvetsov proposed in 1957 that distinctions also be made between progressive and regressive diagenesis. It is perfectly obvious that regressive epigenesis cannot convert a lithified rock into the initial sediments. But it is useful to distinguish between progressive and regressive epigenesis.

According to many authors, the concept "epigenesis" does not include the surface weathering or hypergenesis of A. E. Fersman, although other authors use the term more broadly, including all changes in lithified rocks except metamorphism.

D. S. Sokolov recently [1957] proposed the following principles for delimiting regressive epigenesis and hypergenesis: the weathering of rocks below the water table should be called regressive epigenesis, whereas weathering above the water table should be called hypergenesis of surface weathering.

We have presented a scheme for employing the phases and stages of deposition of sediments and the developmental history of lithification, a scheme taking into account the basic events in the geologic history of the sedimentary formations (Table 1).

It seems to us that the scheme satisfactorily reflects the variations in the basic epochs of "existence" in the development of sedimentary formations.

The stage of incipient metamorphism is intermediate between normal changes in sedimentary rocks during epigenesis and transformations to clearly metamorphic rocks. N. V. Logvinenko has proposed that sedimentary rocks that have been profoundly altered by incipient metamorphism be termed metamorphous rocks, in distinction to metamorphic rocks. In metamorphous rocks, the fragmental minerals are not recrystallized as is the material in the cement of sandy rocks and in the matrix of pelitic rocks. Many authors refer the stage of incipient metamorphism completely to the epigenetic phase.

B. Special Features of Diagenesis and the Lower Boundary of This Zone

Despite the importance of epigenetic processes, the role of diagenesis, i.e., the processes at work during the phase of lithification of the sediments, is of prime interest to lithologists, especially in relation to the variety and complexity of the transformations occurring at that time. Therefore, a study of diagenesis in sediments, in different types of sedimentary rocks, and in mineral deposits, as well as in Recent deposits is very essential.

In one of our papers we briefly touched on the question of redistribution of material during diagenesis and on the problem of the factors determining this redistribution [Teodorovich, 1950]. These points were discussed in more detail by N. M. Strakhov [1953] for subaqueous, chiefly silty-sandy sediments. In this paper Strakhov noted four phases (more precisely, substages) of diagenesis, considering that, on the whole, they embrace, according to Emery and Rittenberg [1952], a thick sequence of sediments (up to 250 m and more?):

1-2) authigenic minerals generally beginning in an oxidizing environment (a thickness of 25-40 cm), and then in a reducing environment (a thickness up to 10 m);

3) redistribution of material in sediments, with the formation of concretions and with cementation (a few tens of meters?);

4) compaction of sediment (lithification), having begun earlier but here occurring along with dehydration of minerals and recrystallization; the significance of both processes (especially lithification) increases in proportion to the effect of diagenesis in the sediment (embraces a thickness interval up to 150-200 m and more).

The first phase — oxidation mineral growth — is confined to the upper film of sediment, from several millimeters to a maximum of 20-40 cm. The second phase — reduction mineral growth — embraces a thickness of sediment of 2-4 m and more (up to 10 m). The reducing environment persists into the third phase — the redistribution of authigenic minerals. The intense activity of bacteria and their ferments is noted only in the first two phases.

In this view of N. M. Strakhov there are two points far from being universal: 1) that the diagenetic stage for all types of subaqueous rocks extends to depths of approximately 250 m (p. 30); and 2) that there is an upper oxidation film in all subaqueous deposits; in many places the oxidation-reduction boundary is actually somewhat above the surface of the sediment (aside from such specific basins as the Black Sea, where the oxidation-reduction boundary is 150-200 m from the surface of the water, i.e., 50-2000 m above the bottom of the sea).

V. D. Lomtadze [1953, 1955] distinguished two principal stages in the lithification of argillaceous sediments: 1) the conversion of argillaceous mud to mudstone, and 2) the conversion of mudstone to argillite. He considered dehydration to be the most important process during diagenesis.

V. D. Lomtadze [1955] described the degree of lithification of argillaceous deposits according to their natural moisture, bulk weight, porosity, and consistency of the muds. He has established five gradations (Table 2).

The author does not indicate on what basis he has disposed all the tested rock samples; they cannot represent a uniform sequential series of changes. Furthermore, it is well known that clays and mudstones may both occur in a single depositional unit.

TABLE 1

Phases and Stages in the Development of Sedimentary Deposits

Phase	I. Syngenesis, or the formation of sedimentary rock		II. Epigenesis, or the transformation of lithified sedimentary rock at relatively low temperatures and pressures in the surface zones of the earth's crust — Katagenesis (normal and regressive sediments)		III. Metamorphism, or the transformation of lithified rocks at high temperatures and pressures		
Stage — Example 1	Sedimentation (= syngenesis of A. E. Fersman)	Diagenesis of sediments	Katagenesis (normal epigenesis)	Progressive epigenesis	Incipient metamorphism	Distinct metamorphism	Complete transformation
Stage — Example 2	Halmyrolysis		Katagenesis	Hypergenesis	—	—	—
Stage — Example 3	Sedimentation	Diagenesis of sediments	Katagenesis	Hypergenesis	—	—	—
Stage — Example 4	Sedimentation	Diagenesis of sediments	Kata-genesis	Hyper-genesis / Kata-genesis	Incipient metamorphism	—	—
Type of sedimentary formation	Sediment	Lithified sediment	Sedimentary rock	Sedimentary rock	Metamorphous sedimentary rock	Paraschists and paragenesis	Paraschists and paragenesis

In a paper in 1956, N. M. Strakhov distinguished only three phases of diagenesis (oxidation of minerals, reduction of minerals, and the reconstitution of authigenic minerals) and he calls the entire stage of diagenesis biogenic (microbic) stage of sediment transformation or of concretion development. Both views are factually imprecise; and, in addition, the third stage of diagenesis of sediments has been described by N. M. Strakhov with no concrete data from study of Recent sediments and rocks, being based purely on theoretical consideration. Moreover, it is generally known [Pustovalov, 1933, 1940; Teodorovich, 1947, 1949, 1954_2, 1956] that all deposits of the sulfide and sulfide-siderite geochemical facies were formed in an environment in which the oxidation-reduction boundary in the initial sediments occurred distinctly above or scarcely above the surface of the sediment. In other words, all the deposits of these facies, according to N. M. Strakhov, went through only two phases of diagenesis: reduction mineral growth and reconstitution. It is quite clear that such a general consideration of diagenesis can furnish lithologists with little that is concrete.

TABLE 2

Gradations of Lithification in Argillaceous Deposits (from V. D. Lomtadze, 1955)

Index	Argillaceous mud	Soft clay	Compact clay	Mudstone	Argillite
Natural moisture, %	75-80	80-30	35-12	15-3	≤ 3-4
Bulk weight of dried rock g/cm³	0.6-0.8	0.8-1.40	1.35-1.90	1.90-2.65	2.65-2.75
Porosity, %	75-80	80-40	45-25	30-4	4-5
Coefficient of porosity.	3-4	3-0.60	0.60-0.35	0.30-0.10	0.10
Consistency.	Fluid of viscous	Plastic	Plastic-semi-solid	Semisolid	Solid

Practically speaking, in his 1956 paper N. M. Strakhov distinguished only two phases of diagenesis in sediment: a first stage of mineral growth and a second phase of reconstitution of authigenic minerals, or the formation of concretions. It is impossible to agree with such contradictions in time between the processes of formation and of reconstitution of authigenic minerals.

According to special studies by A. V. Makedonov, made over a number of years, carbonate concretions form during the early stages of diagenesis of sediments (the early diagenetic stage), and not during the third phase (the stage of late diagenesis), as N. M. Strakhov believes. A. V. Makedonov's data show that the lithification phase in rocks comes after the formation of concretions.

In his 1956 paper, N. M. Strakhov, at least for argillaceous subaqueous sediments, placed the lower boundary of the zone of diagenesis of sediments at a depth of 250-300 m above the floor of the basin, stating that above this depth material is still being redistributed and concretions are being formed. We have shown in our earlier works [Teodorovich, 1952, 1954_1] that carbonate and siliceous deposits (to say nothing of the typical chemical sediments, phosphorites, etc.) become lithified quickly. M. S. Shvetsov noted in 1956 that if we assume the lower boundary of the zone of diagenesis of sediments to lie at a depth of 250-300 m below the basin floor, many of the Paleozoic deposits on the Russian platform must be called sediments.

It should be pointed out that diagenesis is different for different groups of sediments and even for different lithic types of sediments within a single group of sediments that are accumulating and being buried; it also depends on the climatic-geographic zone of accumulation. As we have emphasized [Teodorovich, 1952], a petrographic study of many thin sections shows that limestones, dolomites, and many silicates generally become lithified quickly; this is attested by the nature of organic skeletal remains found in these rocks — undeformed and unflattened. On the other hand, clays preserve their capacity to compress much longer. In clays, marls, siltstones, carbonate-bearing clays, and mudstones, as shown by studies of many thin sections, fossil skeletal remains, spores, and, in places, oolites and other structures become flattened or deformed [Teodorovich, 1954_1].

From the data cited it is quite clear that lithification will occur at markedly different times and at different depths in sediments that are accumulating and being buried and that differ in their fundamental composition and structure. For example, calcareous and siliceous layers will be already compacted and lithified while sandy and, especially, argillaceous sediments continue to be affected by diagenesis (of sediments) for a long time. In any case, many deposits have completed diagenetic changes long before being buried to a depth of 200 m below the floor of the basin, and even argillaceous sediments finish the diagenetic stage much before this. In the experiments of V. D. Lomtadze, which are commonly alluded to [Strakhov, 1956], it was naturally impossible to take into account the role of geologic time. The discontinuity point (zone) on the curve of compaction in clays, equal to 60-80 atm, corresponds to short-period application of pressure on the clays under laboratory conditions.

Frequently the formation of sediments themselves, for example hydrogoethite-chamosite-siderite ores and nodular phosphorites, is rather complex, being accompanied by repeated temporary interruptions and, as a consequence, by changes in the oxidation-reduction conditions and in the rH profile of the sediment.

The basic questions in the study of diagenesis in sediments are the following: the cause of different rates of lithification in the various types of deposits, the cause of changes in porosity in argillaceous (and mixed) deposits in porportion to the accumulation of sediments, the nature (type) of transformation of organic material and the effect of the climatic factor, the real causes of displacement of material during diagenesis, the significance of initial composition of the sediments, the moisture content of the muds, and the changes in moisture content with continued accumulation of sediment.

N. M. Strakhov terms all diagenesis of sediments the biogenic stage; it is difficult to agree with this because the life activity of microorganisms (as is well known) quickly declines in proportion to the cover of subaqueous sediment, and for a number of geochemical facies the role of organic substances during diagenesis is practically nill. It is also impossible to agree with the view that diagenesis is the stage during which concretions develop. The fundamental differences between rocks and sediments are, as is well known, uniform lithification, cementation, and compaction. Lithification during diagenesis of sediment occurs in calcareous, dolomitic,. siliceous, phosphatic, halogen, and, as a rule, silty-sandy deposit, and also in ore accumulations of Fe, Mn, and Al; whereas argillaceous sediments generally experience only compaction during the same interval (only certain types of kaolinitic refractory clays become indurated during diagenesis).

Calcareous deposits of a reef framework or calcareous crusts on the sea floor and other similar formations are lithified practically at the time they are formed. On the other hand, subaqueous argillaceous deposits undoubtedly experience a diagenetic stage to a depth of tens of meters: at places 20-50 m, locally deeper, depending on the composition and structure of the clay muds and on the type and thickness of the sediments interbedded with them and also on the rate of accumulation of the sediments.

Thus, the leading factors in the phase of diagenesis of sediments are the induration and compaction of the deposits. It is impossible to unify the stages of diagenesis for sediments of all types and even for all subaqueous sediments. The normal reference to the presence of ancient friable sands in Paleozoic or other ancient deposits is immaterial in our point of view; during diagenesis sands are generally lithified in any kind of subaqueous basin deposit. If sands are found in Paleozoic or other ancient subaqueous basin deposits, it becomes necessary, after determining the areal distribution, to discover the cause of their lack of consolidation, which, in our opinion, is generally "decementation" (possibly the solution of calcareous cement during the incursion of neutral or acid waters into the rock, perhaps the solution of calcium sulfate cement by freshening of surface waters or by formational waters at depth, etc.).

One cannot deny stages of mineral growth and reconstitution during diagenesis, since they overlap each other (for various chemical combinations). As M. S. Shvetsov maintained in 1956, the mineral matrix of most limestones, microgranular dolomites, silicites, phosphorites (to say nothing of all salt deposits) is primary, not diagenetic.

The role of organic material in the muds during the first stage of diagenesis is considerable, since this constituent regulates the processes of reducing minerals and, with other factors, determines the type of mineral-geochemical facies in the sediment. However, the oxidation-reduction conditions are relatively equalized during the later stages of diagenesis.

In 1957, N. M. Strakhov proposed a third scheme for the sequence of stages of alteration in sediments of subaqueous formation. According to this view a rock is formed during lithogenesis, which consists of two stages, sedimentation and then diagenesis; changes in already lithified rocks belong to the phase of metagenesis. We have noted above the unsuccessful use of the term "metagenesis" in place of the term " epigenesis," since they are synonymous.

The nature of diagenesis, according to these later views of N. M. Strakhov, is found in the development of processes that eliminate internal inequilibrium in the sediments, that lead to the formation of rock. In principle these may also be no more than biogenic processes, though they cannot be all the processes bringing a sedimentary system (sediments — muddy water) to physicochemical equilibrium. In other words, even this statement of the problem does not demonstrate the complete biogenic nature of diagenesis even for certain subaqueous sediments in humid zones that accumulated during the initial stage of increasing salinity in a basin. But there is a whole series of other types of sediment accumulations (arid, semiarid, volcanic-sedimentary, etc.), for which, as well as for most continental subaerial deposits, one cannot admit that the entire diagenetic stage as a whole is biogenic. According to N. M. Strakhov (and as assumed by lithologists previously), the conditions of diagenesis are very similar in temperature and pressure to the conditions of sedimentation. However, this view provides us with no criterion for drawing a lower boundary for the zone of diagenesis or for the boundary between diagenesis and epigenesis. The temperature in the zone of diagenesis of sediments at any assumed thickness will differ little from the temperature in the zone of sedimentation, but the pressure will be markedly different, and it is not known what the value may be at the boundary between the zones of diagenesis of sediments and epigenesis (Table 3).

TABLE 3

Different Views on the Thickness of the Zone of Diagenesis

Author	Depth to the lower boundary of the zone of diagenesis below the basin floor, in m	Maximum difference (average)	
		for temperature	for pressure
L. B. Rukhin	15-20	0.5°	6 atm
M. S. Shvetsov. . .	30-50	1.5°	15 "
N. M. Strakhov. . .	100-200-300	9.0°	80 "

The physical-mechanical properties of the rocks are not considered in this scheme.

In addition, all the discussed calculation are chiefly theoretical, since it was shown by us above that different types of deposits are converted into rocks at various depths; one may thus speak about the lower limit of the diagenetic zone only for deposits of a given type or of a given group of types.

According to N. M. Strakhov, three phases of diagenesis may be distinguished: 1) oxidation, 2) reduction, 3) reconstitution of material.

In connection with these three phases it is necessary to state the following:

1) the phase of oxidizing mineral growth is commonly absent in marine and other sediments (sulfide, sulfidesiderite, and siderite or chamosite mineral-geochemical facies);

2) the reduction phase is absent in all subaerial deposits and in subaqueous deposits of all types of oxidizing facies;

3) the reconstitution phase may occur after the oxidizing and reducing phases, but it commonly occurs simultaneously with the latter one; it is therefore artificial to indicate a break in time between the second and third phases. For example, according to A. V. Makedonov the concretions in the Lower Permain coal-bearing sequence of the Pechora basin formed during the stage of early diagenesis of sediments, and not in the later part of this stage. N. M. Strakhov maintains that all colloidal induration occurs sporadically, but this is a special case; most commonly induration progresses throughout the entire mass more or less uniformly (patchiness of induration is found in the formation of concretions; it does not refer to rocks that are indurated with the development of macroconcretions).

According to N. M. Strakhov [1957], the diagenesis of sediments signifies chiefly changes in mineral composition and in redistribution of material during rather weakly expressed induration, whereas during epigenesis the physicochemical properties change markedly (induration increases) and there is little change in the mineral and structure-textural properties. This view is generally valid only for changes in the mineral composition.

Epigenesis, or metagenesis according to N. M. Strakhov, is characterized by the fact that the rocks are depressed into zones of ever increasing temperature and pressure; the equilibrium of the system is disturbed and processes of adjustment begin to work. At this phase two stages are distinguished — epigenesis proper (or katagenesis) and incipient metamorphism; the later stage may be called the stage of eometamorphism, but the grounds for referring incipient metamorphism to epigenesis (metagenesis) are not clear, since the term better represents an independent stage. During epigenesis, pressure changes markedly, temperature rather slightly.

From data on platform sections, N. M. Strakhov has assumed the lower boundary of epigenesis to occur at a depth of about 3000-3500 m (perhaps deeper), i.e., at a pressure up to 800-900 atm and a temperature up to 100°. The stage of incipient metamorphism is typically developed in geosynclinal regions.

It should be noted that the change in physicochemical conditions in a rock depends not only on depth. For example, mudstones and clays are commonly found in a single unit at depths of about 1500 m. This fact points up the effect of initial composition of the deposits and of several other factors (apart from depth of burial) on the degree of compaction and lithification.

During epigenesis successive changes in types of subsurface waters with depth may be significant: bicarbonate→sulfate→chloride. However, we also have data on changes in type of water with various values of pH and rH, which show no definite trend. The study of the processes of epigenesis and the discrimination of the boundaries between the zones of epigenesis, incipient metamorphism, and typical regional metamorphism represent the current problem facing lithologists; the first data concerning work of this kind have been obtained during a study of Mesozoic and Paleozoic rocks in the Verkhoyansk region and of Middle Carboniferous strata in the Donets Basin.

Chapter III

AUTHIGENIC, PRINCIPALLY SYNGENETIC MINERALS IN SEDIMENTARY ROCKS, THEIR CHARACTERISTICS, CONDITIONS OF FORMATION, AND CLASSIFICATION

We shall not take space here to describe minerals present in sedimentary rocks as fragmental grains and derived from weathered igneous or metamorphic rocks (clastic quartz and feldspars, micas, amphiboles, pyroxenes, etc.).

Among the authigenic sedimentary minerals we can describe only the more or less widespread groups and species.

Authigenic minerals in sedimentary rocks may be found in the following forms: a) clearly crystalline anisotropic grains, each visible in thin section, affecting polarized light and not covered by other grains; b) microcrystalline anisotropic grains, of such a size (d ≤ 0.01 mm) that they lie one above the other within the thickness of a single thin section; c) colloidal material, particles smaller than 0.1 μ, which always appear isotropic in thin section although they may be amorphous and crystalline; d) distinctly crystalline and microcrystalline isotropic grains, rarely abundant (fluorite, etc.); e) clearly amorphous, i.e., noncrystalline and isotropic; and f) distinctive and, in places, very characteristic microaggregates.

At the present time, when occasion demands, the authigenic minerals in sedimentary rocks may be studied by a whole series of methods:

1) in thin sections with a polarizing microscope or in polished sections with an opaque illuminator;

2) in grains under a binocular microscope or in immersions under the polarizing microscope;

3) by using staining reagents on uncovered thin sections, polished sections, polished thin sections, etc.; by etching polished sections and polished thin sections; and also by using a system of dyes specially designed for the investigation of preparations of the clay-group minerals;

4) by microchemical analysis;

5) by thermal analysis, heating curves, and cooling curves;

6) by x-ray examinations, analysis of x-ray powder photographs and comparison with standards;

7) by use of the electron microscope;

8) by means of the blow pipe or by use of individual characteristic qualitative reactions;

9) by chemical analysis;

10) by spectral analysis.

Most of the work by lithologists has to do with the first three methods. The present paper will describe the principal diagnostic features obtained by these three methods. Mineral characteristics obtained by using the remaining methods of investigation are given where they are very diagnostic or where it may be impossible to identify the minerals by use of the first three methods alone.

We shall describe only the principal authigenic minerals in sedimentary rocks, the most widespread of which are the minerals in the silica, carbonate, and silicate groups; less widespread groups are the oxides and hydroxides of aluminum, iron, manganese, and copper, the sulfides of iron and copper, the phosphates, the sulfates and fluorides, the soluble salts, and the native elements.

In contrast to the normal description of minerals in keeping with the chemical classification, i.e., beginning with the native elements and ending with the silicates, we shall describe the authigenic minerals in sedimentary rocks according to chemically homogeneous groups, but beginning with the more widespread forms and ending with the rarer ones. The mineral descriptions are introduced in the order outlined in detail in the table of contents.

1. SILICA GROUP

Opal — $SiO_2 \cdot nH_2O$, or $SiO_2 \cdot aq.$ A colloidal mineral (solid hydrogel), colorless or stained various shades by impurities. The hardness is 5.0-6.0; the specific gravity 1.9-2.3; water content ranges from 1-2 to 5-15, rarely up to 34%; hardness and specific gravity increase with decrease in water content (the presence of adsorbed heavy ions increases with specific gravity). On heating, some varieties of opal lose most of their water at a temperature below 100°, others between 100 and 250°.

X-ray studies have shown that sedimentary low-temperature opals, apart from the chief mass of x-ray-amorphous silica, belong partly (the dispersed phase) to alpha-quartz [Betekhtin, 1950], or, more commonly, to low-temperature cristobalite: alpha-cristobalite [Bushinskii and Frank-Kamenetskii, 1954]. Silica that is amorphous to x-rays forms completely separate siliceous rocks, some diatomites for example; organic remains also constitute several other silicites, occurring with dispersed components that are anisotropic to x-rays.

Opal is optically isotropic, but because of internal strain may be weakly birefringent. The refractive indices range from 1.26-1.40 to 1.46-1.48 (most commonly about 1.43-1.45), i.e., always lower than Canada balsam. The indices decrease with increase in water content; at 8.97% water the index is 1.447, but at 3.55% it is 1.459.

The very low index of refraction in opal produces a negative shagreen surface on the mineral.

Opal in sedimentary rocks is found in compact masses (in opaline clays) or as cement of homogeneous gelatin-like opal (in some sandstones), in small spheres or globules ranging from 0.003-0.005 to 0.03-0.05 mm across, in globular tripoli (Fig. 1), in skeletal remnants of siliceous organisms (sponge spicules, and fossil diatom

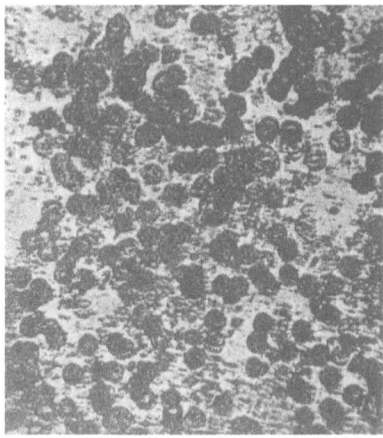

Fig. 1. Tripoli, consisting of opaline globules. × 225 (From Cayeux, 1916).

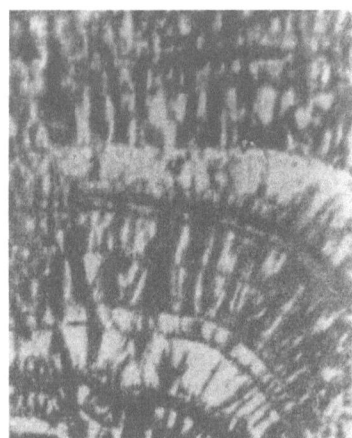

Fig. 2. Alternating zones of quartzine (a) and chalcedony. × 45, nicols crossed. (From Cayeux, 1916.)

shells and radiolarian skeletons, which apparently consisted of organic silica compounds during the life of the organisms), and, finally, in varieties of sinter.

By recrystallizing and losing water, opal grades into minerals of the chalcedony group, and then into quartz. Several investigators have pointed out that the presence of $CaCO_3$ greatly accelerates the recrystallization of opal. The hardness, specific gravity, and index of refraction increase with dehydration.

In sedimentary deposits opal may be clearly of organic origin (skeletons and shells of organisms), may be dissolved and be precipitated from solutions, may be a chemically colloidal sediment, or may have a volcanic-sedimentary origin; it may also be widely developed in the zone of weathering. Opal is easily soluble in an alkaline environment, and is identified in rocks by extraction with a 5% solution of soda or potash or a 5% solution of KOH.

Diatomites, tripoli (globular), opaline clays, geyserite, and other rocks belong to the group of opaline rocks.

Chalcedony—SiO_2. The combined term for all fibrous (i.e., structural) varieties of anhydrous crystalline silica (according to a number of authors it is entirely quartz): quartz, lutecite, chalcedonite, and pseudochalcedony (the diagnostic features are given in Table 4); some authors use the term "chalcedony" in a more restricted sense, as a synonym for chalcedonite. The hardness of chalcedony ranges from 6.5 to 7.0; the specific gravity is 2.5-2.65. The crystal system has not yet been precisely established, but, in any case, chalcedony is found in microcrystalline and cryptocrystalline masses, in normal-fibrous incrusting segregations or in radial microaggregates, and also in the skeletal remains of siliceous sponge spicules and radiolarians, having been recrystallized in this later occurrence. Most authors consider chalcedony to be a fine, fibrous variety of quartz, containing admixtures of opal and submicroscopic cavities. X-ray examinations show the minerals of the chalcedony group to be cryptocrystalline derivatives of alpha-quartz.

TABLE 4

Optical Properties of Fibrous Varieties of Crystalline Silica and Quartz

Mineral	Sp. gr.	System or No. of optic axes	Highest index of refraction	Lowest index of refraction	Birefringence	Optical character mineral	Optical character elongation	Extinction
Quartz.....	2.65	Trigonal	1.553	1.544	0.009	(+)	(+)	Parallel
Quartz.....	2.55	(?)	1.544	1.533	0.009-0.011	(+)	(+)	Parallel
Lutecite....	2.55	Biaxial	1.543	1.533	0.010-0.009	(+)		Inclined $c^{\wedge}Ng = = 29\text{-}30°$
Chalcedonite.	2.5-2.6	Biaxial (?)	1.538-1.547	1.530-1.537	0.008-0.010	(+)	(−)	Parallel
Pseudochalcedony.....	2 5	Biaxial (?)			0.0045	(−)	(−)	Parallel

Normally, minerals of the chalcedony group are derived from recrystallization of opal, and there may be gradations between the two, but chalcedony minerals may also, apparently, precipitate directly from solution. After being recrystallized these minerals gradually change into quartz with time, the closest approach to which is quartzine (Fig. 2). Minerals of this group are easily distinguished in polished sections from sometimes similar gypsum because they are not etched by 2% HCl and because they are unaffected by long washing in water.

Minerals of the chalcedony group are characteristic of jasper and, in general, of radiolarian crystalline silicites, many sponge silicites (spongolites), spicule-rich silicites (in which the chalcedony commonly preserves the form of the opaline globules), many cherty limestones, and siliceous formations in carbonate sequences.

We should note lussatite here, a cryptocrystalline and finely fibrous modification of silica containing up to 8% water (some authors think this to be due to an admixture of opal), with a specific gravity of 2.04-2.27, and refractive index Nm of about 1.447-1.467. The mineral is optically positive, has positive elongation, and has a birefringence of about 0.004.

It is distinguished from quartz and quartzine by smaller birefringence and by the water content. Some authors think lussatite is an unstable variety of low-temperature cristobalite, or alpha-cristobalite (a fact now confirmed by x-ray studies), and, besides, to consist partly of amorphous silica (x-ray-amorphous opal).

Lussatite proper possesses positive elongation; if the elongation is negative the mineral is called lussatine. These minerals belong to the opal series: lussatite generally forms during the syngenetic stage or during early diagenesis of sediments (V. S. Vasil'ev).

Quartz — SiO_2

Distinctly crystalline (nonfibrous) anhydrous silica. It crystallizes in the trigonal system. In well-formed crystals two rhombohedrons are developed $(10\bar{1}1)$ and $(01\bar{1}1)$, and the prism $(10\bar{1}0)$. The hardness is 7, the specific gravity 2.65-2.66. It is colorless, rarely stained with impurities (in thin section it is colorless).

The following varieties of quartz may be distinguished according to the color of transparent crystals: 1) rock crystal — colorless crystals; 2) amethyst — a violet variety; 3) morion — black crystals of quartz; 4) citrine — golden yellow or lemon yellow crystals; 5) smoky quartz — crystals of quartz with grayish or brownish shades. Cleavage in quartz is practically absent.

The indices of refraction are Ne = 1.5533 and No = 1.5442; Ne — No = 0.009; the mineral is uniaxial, positive; prismatic sections give parallel extinction, and the elongation is positive. Quartz grains are most widespread among clastic particles, but the authigenic varieties are what interest us here. Normal sedimentary quartz (trigonal) is alpha-quartz, in contrast to hexagonal high-temperature beta-quartz.

Authigenic quartz in sedimentary rocks is found in characteristic idiomorphic (Fig. 3) or irregular grains, in fine-grained and very fine-grained masses, as pseudomorphs after sponge spicules (initially composed of opal and then chalcedony) and rarely of radiolarian skeletons, as enlargements (overgrowths) about fragmental grains (having the same optical orientation) (Fig. 4) rarely in characteristic microaggregates (such as radiolites, veinlets, generally products of recrystallization of incrustations of fibrous varieties of silica), and in individual crystals or druses in pores and cavities in rocks.

During the enlargement of fragmental quartz grains a mosaic of quartz grains is developed, characteristic of quartzites and quartzitic sandstones (Fig. 4). At first glance at a thin section it appears as if there is no cement, but careful examination reveals a film on the surface of the clastic grains or discloses small inclusions in the grain that contrast with the clear overgrown border of authigenic quartz.

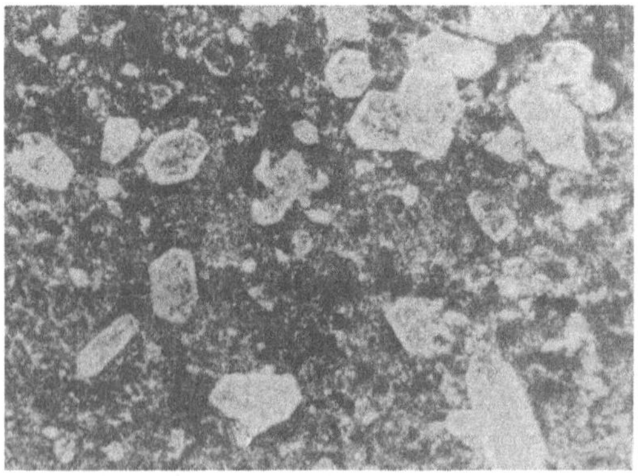

Authigenic quartz in sedimentary rocks forms during late diagenesis and during epigenesis, and is generally the product of recrystallization of opal and minerals of the chalcedony group, though it sometimes precipitates directly from weakly mineralized waters during their slow advance. In geodes of the Middle and Upper Carboniferous sedimentary rocks, for example, one may occasionally find rock crystal, citrine, and amethyst (the village of Rusavkino in the Moscow oblast), formed by precipitation from circulating formational water or from surface water of atmospheric origin (vadose water, not deep water).

Fig. 3. Idiomorphic grains of authigenic quartz in limestone. × 15. (From Cayeux, 1916.)

Quartz, and chalcedony and opal of sedimentary rocks even more, may be dissolved in the layers of sedimentary rock and may be replaced by sulfides and hydroxides of iron, by siderite, by glauconite, occasionally by calcite, by manganese hydroxides, and by other compounds. This fact is clearly borne out by the presence of corroded quartz grains and by unquestionable replacement of the quartz by other, authigenic minerals during changes in the physicochemical environment.

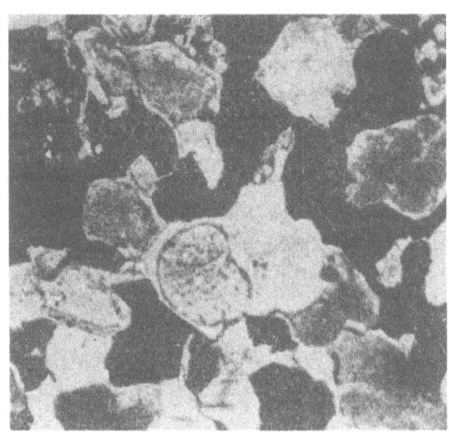

Fig. 4. Enlargements (overgrowths) of clastic quartz grains at the expense of siliceous cement, grading into the quartz and joining the quartz grain with the same optical orientation. × 25; nicols crossed. (From Cayeux, 1916.)

Authigenic quartz is found in many siliceous rocks (phthanites, jasper, radiolarian and spongian silicites), as cement in sandstones (especially in quartzites and quartzitic sandstones), and also in disseminated grains, aggregates, and zones of rock in cherty limestones and in some limestones and dolomites.

2. CARBONATE GROUP

Carbonates, or the salts of carbonic acid (H_2CO_3), are typical authigenic minerals in sedimentary rocks, forming the principal component in rocks or as basal cement* during syngenesis (sedimentation and diagenesis of sediments); in subordinate amounts or in disseminated forms they develop during syngenesis (diagenesis) and epigenesis. Carbonate minerals are chiefly distinctive of sedimentary rocks, but they may also be of hydrothermal origin. Among the anhydrous carbonate minerals, two chief subgroups are distinguished: the calcite subgroup, carbonates in the trigonal system, and the aragonite subgroup, carbonates in the orthorhombic system.

In addition to the five principal minerals — calcite $CaCO_3$, dolomite $CaMg(CO_3)_2$, siderite $FeCO_3$, magnesite $MgCO_3$, and rhodochrosite $MnCO_3$ — the calcite subgroup also contains ankerite and several series of isomorphous mixtures: siderite — magnesite, siderite—rhodochrosite, and rhodochrosite—calcite. The siderite—calcite series is apparently discontinuous in the middle:

Siderite — $FeCO_3$ from 100 to 95%; $CaCO_3$ from 0 to 5%;
Siderot — $FeCO_3$ from 95 to 80%; $CaCO_3$ from 5 to 20%;

. . . .

Ferrocalcite — $FeCO_3$ from 15-10 to 1%; $CaCO_3$ from 85-90 to 99%;
Calcite — $FeCO_3$ from 1 to 0%; $CaCO_3$ from 99 to 100%.

The aragonite subgroup, or the family of orthorhombic carbonates, contains the following minerals that are known in sedimentary rocks: aragonite $CaCO_3$, strontianite $SrCO_3$, witherite $BaCO_3$, and cerussite $PbCO_3$.

Basic carbonates known in sedimentary deposits are malachite $Cu_2(CO_3)(OH)_2$ and azurite $Cu_3(CO_3)_2(OH)_2$.

Finally, hydrous and chiefly hydrous carbonates known in sedimentary rocks include natron (soda) $Na_2CO_3 \cdot 10H_2O$, thermonatrite $Na_2CO_3 \cdot H_2O$, trona $Na_3H(CO_3)_2 \cdot 2H_2O$, hydromagnesite $Mg_5(CO_3)_4(OH)_2 \cdot 4H_2O$, artinite $Mg_2(CO_3)(OH)_2 \cdot 3H_2O$, nesquehonite $MgCO_3 \cdot 3H_2O$, and lansfordite $MgCO_3 \cdot 5H_2O$. Of these, we have described only natron, hydromagnesite, and nesquehonite as being relatively more abundant in sedimentary rocks.

The most widespread carbonates in sedimentary rocks (as authigenic minerals) are calcite and dolomite, followed by siderite, magnesite, rhodochrosite, and ankerite.

Anhydrous Carbonates

Calcite Subgroup

Calcite—$CaCO_3$. A very widespread rock-forming carbonate mineral of the trigonal system, also called calcspar. It normally contains very small admixtures of Mg, Fe, Mn, and, rarely, other carbonates. Chemically pure calcite consists of 56% CaO and 44% CO_2. Calcite with a small content of isomorphous admixture of $FeCO_3$ (from 1 to 10-15%) is called ferrocalcite. The hardness of calcite is 3, the specific gravity 2.6-2.8 (2.71-2.72 for very pure material). Calcite crystals, found only in cavities, have a variety of forms: most frequently scalenohedral, more rarely tabular, prismatic, or rhombohedral. It is mostly colorless or milk white, but it may take on a number of shades because of impurities. Perfect rhombohedral cleavage is characteristic, and large grains

---

*Tr. note: Basal cement is a type of cement in sedimentary rocks generally exceeding in volume the clastic grains; a distinctive feature is the failure of clastic grains to be in contact with each other.

21

commonly show polysynthetic twinning. The refractive indices are No = 1.6585 and Ne = 1.4853; No−Ne = 0.1722. It is uniaxial, negative, and possesses a pseudoabsorption (i.e., in sections near the principal section, the sha-green surface appears and disappears on rotation of the stage).

The mineral is found in sedimentary rocks chiefly in fine-grained, distinctly granular, or microgranular masses and as cement in clastic rocks, and, in addition, it forms most of the skeletal remains of organisms (for-aminifers, echinoids, ostracods, trilobites, bryozoans, articulated brachiopods, corals, and pelecypods (in part), and calcareous algae); the mineral is also found in calcareous incrustations and oolites or as products of the life activity of blue-green algae.

The principal structural-genetic types of sedimentary chemically formed $CaCO_3$ (calcite chiefly) are a) microgranular, b) calcareous crusts, and incrustations, c) concentric layers in oolites, d) diagenetic (syngenetic) recrystallization, e) crystallized coatings on the walls of cavities, and f) epigenetic recrystallization and filling of cavities.

Calcite is dissolved in 2 to 5-10% hydrochloric acid with effervescence, even when the acid is cold; it also dissolves in carbonic acid. The easy solubility of calcite at low pH values leads to its replacement by silica, dolomite (especially at high values of pCO_2), and other minerals; occasionally calcite, being precipitated from alkaline solutions, replaces these other minerals, even quartz.

Calcite is chiefly of chemical origin; it precipitates in microgranular varieties (sometimes as incrustations or oolites) which may later be more or less recrystallized and become coarser grained. Calcite of organic origin is also of great importance. Initially $CaCO_3$ may be precipitated in a colloidal variety, bütschliite; this form is unknown in rocks because it quickly changes to metacolloidal calcite. Moreover, even aragonite changes to calcite in time (for example, shells of gastropods, pteropods, and, in part, pelecypods).

Among the sedimentary rocks, calcite forms limestones and the main parts of marls and dolomitic lime-stones; it forms all or much of the cement in many sandstones, siltstones, and conglomerates; it is present in disseminated form in many argillaceous rocks (calcareous clays and corresponding mudstones). Calcite is also found in lenses of secondary distinctly crystallized limestones, representing the products of dolomite alteration (dedolomitization).

Dolomite − $CaMg(CO_3)_2$. A widespread rock-forming mineral, consisting of the double salt of $CaCO_3$ and $MgCO_3$. Chemically it is 30.4% CaO, 21.7% MgO, and 47.9% CO_2. Dolomite crystallizes in the trigonal system, normally in rhombohedral grains. The hardness is 3.5-4.0, the specific gravity 2.8-2.9. The color is brownish white or yellowish. Cleavage is also the rhombohedron. The indices of refraction are No = 1.680-1.682, and Ne = 1.500-1.502; No−Ne = 0.179-0.182; the mineral is uniaxial, negative, and has pseudoabsorption. Iso-morphous admixtures of Fe^{+2} and Mn^{+2} increase the refractive indices: No to 1.695 and more, Ne to 1.513 and more.

Dolomite forms pelitomorphic or microgranular masses (Fig. 5) or cement in rock, fine-grained masses or cement of more or less equant rhombohedrons (Fig. 6), distinctly grained masses of irregular rhombohedral grains, idiomorphic rhombohedrons in calcite masses in dolomitic limestones, occasional crusts and incrustations, oolites, and similar formations. Rhombohedrons of dolomite commonly have a central nucleus (of micro-granular carbonate, clay, or iron compounds) or a zoned structure (Fig. 7). The chief structural-genetic types of sedimentary dolomite are a) pelito-morphic − of irregular grains chemically precipitated from waters in the basin; b) replacement − of relatively coarser rhombohedral, rounded-rhombohedral, and similar grains; c) overgrowths on the walls of cavities (of hemispherical or sheaf-like microgranular aggregates, showing extinc-tion crosses); d) incrustations; e) oolitic concrentric growths; f) distinctly crystallyine recrystallized zones, generally forming as a result of incipi-ent metamorphism of the rock; and g) rarely, axiolites, radiolites, and spherulites.

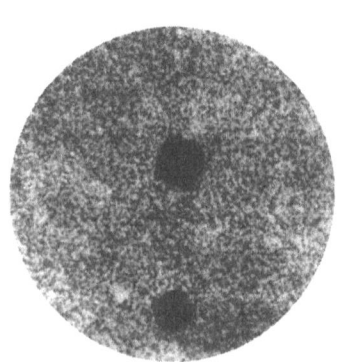

Fig. 5. Pelitomorphic dolomite with cavities due to leaching. ×85; nicols crossed. (From Pustovalov, 1937).

Dolomite dissolves in hydrochloric acid much more slowly, i.e., less energetically, than calcite. On a piece of rock effervescence is scarcely noticeable, but on powder effervescence is rather marked. Dolomite dissolves in 10% hydrochloric acid when heated. Staining reactions are used to distinguish dolomite from calcite in powder form or in uncovered thin sections, calcite being stained and dolomite remaining unstained or almost unstained: there are three Lemberg reactions — with $FeCl_3$, $AgNO_3$, and logwood extract [Teodorovich,1950, p. 35]; a reaction with copper nitrate [Tatarskii, 1955, pp. 45-46]; and staining with acidified violet inks [idem]. To distinguish calcite from dolomite in polished sections, the material is etched in 2% hydrochloric acid for 20 seconds (calcite becomes etched; dolomite remains polished); this technique was proposed by D. S. Belyankin, V. V. Lapin, and I. A. Ostrovskii [1940]. On thermal curves (in contrast to calcite, which is characterized by a single endothermic reaction at 900-930°), dolomite shows two endothermic effects (at 700-770° and at 900-930°); even a very small amount of alkaline-metal salts in the material will lower the temperature of the first effect.

We give a brief description of the chromatic reactions for distinguishing dolomite from calcite in thin sections. All the staining reactions on uncovered thin sections give better results when filter paper, previously moistened in distilled water, is used.

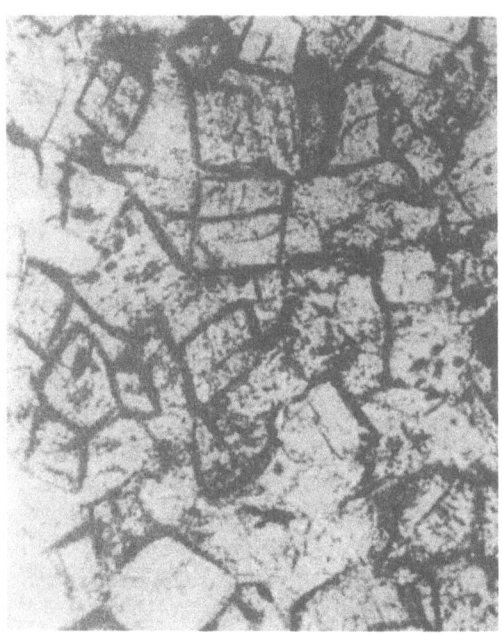

Fig. 6. Extremely fine-grained rhombohedral structure of a dolomitic rock. × 180. (From Cayeux, 1916.)

The first Lemberg reaction involves the treatment of the thin-section surface, or a part of it, with 10% Fe_2Co_6 for 1-2 minutes; this will produce a reddish brown film of iron hydroxide, $Fe(OH)_3$, on the surfaces of calcite grains. Then, after washing, the surface of the thin section is treated with a saturated solution of $(NH_4)_2S$ for several seconds. The black coating of iron sulfide that forms may be fixed by a solution of hyposulfite, and the thin section may then be covered again. The solution of Fe_2Cl_6, as well as the solution of $(NH_4)_2S$, is best placed on the wet filter paper with a glass rod.

The second Lemberg reaction utilizes a solution of 60 parts water, 4 parts anhydrous aluminum chloride (Al_2Cl_6), and 6 parts logwood extract (Haemotoxylon campechianum); the solution is then boiled for 25 minutes, the volume being held, by periodic additions of water, at a constant level. When the solution has cooled, a deep violet fluid is filtered off; best results are obtained by using a freshly prepared solution. On the part of the thin section placed in the cold solution, calcite acquires a violet color after 5-10 minutes (dolomite does not change color when left in solution for 20 minutes).

The third Lemberg reaction, with $AgNO_3$, is more conveniently utilized in its second variant; in this technique K_2CrO_4 is used in place of pyrogallic acid after the thin section has been treated with 10% solution of $AgNO_3$ (70°) for 3-4 minutes. After the thin section is removed from the solution of $AgNO_3$ it is immersed for one minute in a 20% solution of K_2CrO_4; during this procedure the grains of calcite acquire a reddish brown or brown-red color.

The reaction with copper nitrate was used by Maler for powders of carbonate material. On boiling in a weak solution of $Cu(NO_3)_2$ for several minutes, calcite powder is stained a bright green color, because of the formation of the chief green copper carbonate. Later [Rodgers, 1940] this reaction, in a modified form, came to be used for staining calcareous-dolomitic rocks in polished sections. A polished section of rock, with the polished side down, is placed for 5-6 hours in a molar solution of $Cu(NO_3)_2$ (corresponding to the content of 188 g of $Cu(NO_3)_2$ in one liter of water) (the polished surface should not rest directly on the floor of the vessel); during this treatment grains of calcite acquire a light yellow color, a film that is easily rubbed off. The polished section is then taken from the solution and immediately, without wiping or washing, placed for several seconds in a strong solution of ammonia, after which the color of the calcite becomes blue and permanent. After this the polished section is washed in water and carefully wiped to ensure complete removal of all $Cu(NO_3)_2$ and ammonia solution. Dolomite does not change color during this treatment, but it does lose its polish. A. M. Gabril'-yan found ways of using this reaction in ordinary uncovered thin sections.

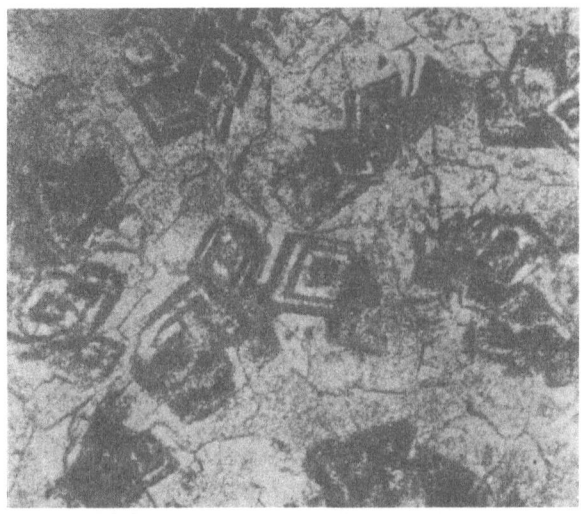

Fig. 7. Small rhombohedrons of dolomite with zonal structure in dolomitic limestone. × 70. (From Cayeux, 1916.)

The use of acidified violet inks gives good results on relatively coarse-grained calcareous dolomitic rocks, the ink apparently being absorbed by finely dispersed clay particles in the calcite. Ordinary (methyl violet) inks, when a small quantity of hydrochloric acid is added, become bright green; on the surface of calcite, these inks become violet again through neutralization of the acid. It is recommended (S. V. Tikhomirov) that the violet inks be acidified with 5% hydrochloric acid in order to obtain a dark blue color. An uncovered thin section is submerged in blue ink for 40-90 seconds and a filter paper is carefully applied; the calcite reacts with the solution and is stained a violet color, but dolomite remains unstained. The film of color on the calcite is impermanent (it is washed off with water and soap).

Sedimentary dolomite may be a primary chemical precipitate (its formation is favored by high partial pressure of CO_2, somewhat increased salinity, and other factors) or it may form by replacement of $CaCO_3$, generally during diagenesis of sediments. In this latter occurrence, the $MgCO_3$ may be derived a) from skeletal remains of organisms of the calcareous-magnesian group, b) from sea water in slow, warm bottom currents, c) as a result of reaction between $(NH_4)_2CO_3$ (from the decomposition of organic material) and $MgCl_2$ in the sea water, d) from increased salinity, by means of Haidinger's reaction ($2CaCO_3 + MgSO_4 \rightleftharpoons CaCO_3 \cdot MgCO_3 + CaSO_4$). Some authors think it probable that calcite and the principal magnesium carbonate salts precipitate chemically from the waters in the basin, at times, and that dolomite is subsequently formed in the sediments during diagenesis.

Most lithologists believe that epigenetic dolomite, having formed in lithified rocks by means of circulating solutions, has a limited solubility.

The mineral dolomite forms various types of dolomites and calcareous dolomites, the cement in many siltstones and sandstones (in whole or in part), and is an important constituent in all dolomitic limestones and marls, dolomitic sulfate rocks, and dolomitic clays and mudstones.

Ankerite — $Ca(Mg, Fe)(CO_3)_2$. Ankerite is a variety of calcium and magnesium carbonate enriched in $FeCO_3$ in the molecular composition. $MnCO_3$ is frequently present in considerable quantities, and the formula is then $Ca(Mg, Fe, Mn)(CO_3)_2$.

The chemical composition of normal, or typical, ankerite may also be expressed as $2CaCO_3 \cdot MgCO_3 \cdot FeCO_3$; however, the ratio of $MgCO_3$ to $FeCO_3$ is variable, FeO often being more abundant than MgO. Some authors, considering ankerite to be dolomite with isomorphous admixtures of $FeCO_3$ or to be different isomorphous mixtures of dolomite and hypothetical ferrodolomite with a composition of $CaFe(CO_3)_2$, call ankerite iron dolomite. The system is trigonal. The hardness is about 3.5, the specific gravity 2.9-3.1. The color is yellowish, white, gray, brownish.

The mineral is uniaxial and negative. The indices of refraction vary with the contents of Mg and Fe: at Mg : Fe = 1 : 1 (10.07% MgO and 17.94% FeO), i.e., in typical ankerite, No = 1.721; when Mg : Fe = 1 : 2,

No = 1.735; when Mg : Fe = 1 : 3 (FeO + MnO = 25.7% and MgO = 4.8%), No = 1.741, Ne = 1.536, and No−Ne = = 0.205; when Mg : Fe = 4 : 1 (FeO = 7.5% and MgO = 16.9%), No = 1.698, Ne = 1.513, and No−Ne = 0.185.

According to A. N. Winchell and H. Winchell, $CaMg(CO_3)_2$, ordinary dolomite, or magnesiodolomite, apparently forms an isomorphous series in all proportions with $CaFe(CO_3)_2$, i.e., ferrodolomite, and probably also with $CaMn(CO_3)_2$, or mangandolomite (Table 5). These authors call the intermediate members between dolomite and $CaFe(CO_3)_2$ parankerite when Mg : Fe is near 2 : 1 and ankerite when Mg : Fe is near 1 : 1. This usage is problematical since no mineral with the composition of $CaFe(CO_3)_2$ occurs in nature, and there is a considerable gap in the middle of the isomorphous series $CaCO_3−FeCO_3$ (see p. 21); ankerite containing more than 40% $FeCO_3$ or $FeCO_3 + MnCO_3$ is hardly ever encountered.

TABLE 5

Properties and Composition of Minerals in Isomorphous Mixtures of Dolomite with "Ferrodolomite" or Mangandolomite (from A. N. Winchell and H. Winchell, 1953)

Initial components and properties	Magnesiodolomite (dolomite)		Parankerite	Ankerite		Ferrodolomite *	Mangandolomite
$CaMg(CO_3)_2$	100.0	90.0	80.0	54.4	24.7	0.0	0.0
$CaFe(CO_3)_2$	0.0	8.0	20.0	45.6	70.6	100.0	0.0
$CaMn(CO_3)_2$	0.0	0.0	0.0	2.0	4.6	0.0	100.0
No	1.679	1.686	1.698	1.728	1.741	1.765	1.743
Ne	1.502	1.505	1.513	1.531	1.536	1.555	1.546
No—Ne	0.177	0.181	0.185	0.197	0.205	0.210	0.197
Sp. gr.	2.87	2.94	?	3.02	3.12	3.2	3.1

* The properties of pure ferrodolomite were computed.

A. G. Betekhtin [1950] used the term ankerite broadly: for minerals with an Fe : Mg ratio ranging from 3 : 1 to 1 : 4.

Varieties of the ankerite subgroup with distinctly more Mg than Fe were called parankerite by A. N. Winchell and H. Winchell, as noted previously (although it would be more suitable to call them magnesioankerite), whereas ankerite proper was considered to be a mineral with an Fe : Mg ratio of approximately 1 : 1. For varieties of the ankerite subgroup containing clearly more Fe than Mg, the special term ferroankerite should be introduced Furthermore, it is impossible to ignore completely the possible variations in the content of $CaCO_3$. Apparently ankerite may be considered an isomorphous mixture of $CaCO_3$, $MgCO_3$, $FeCO_3$, and, partly, $MnCO_3$, or, more frequently, as dolomite with isomorphous admixtures of $FeCO_3$ and, in part, $MnCO_3$, i.e., as a series of minerals with the general formula $CaCO_3 \cdot (Mg, Fe)CO_3$.

The ankerite minerals are found in a variety of forms in sedimentary rocks: fine-grained and very fine-grained masses in layers and concretions or as cement in siltstones and sandstones, in scattered rhombohedral grains in argillaceous rocks, frequently in microspherulites, radiolites, and similar formations.

Ankerite dissolves in hot 10% hydrochloric acid; it does not effervesce in cold acid (on a piece of rock). It is distinguished from dolomite by higher indices of refraction.

Some data on staining ankerite powders have been proposed by N. V. Logvinenko and N. K. Zabolotnaya [1954]. These authors have shown that ankerite is not stained by any organic dye, and that when treated with inorganic reagents it is stained only by $K_3[Fe(CN)_6]$. Only slight changes in color are noted when using other reagents.

Reaction with potassium ferricyanide. A thin section or grain of carbonate mineral is soaked for 20-30 seconds in a 20% solution of $K_3[Fe(CN)_6]$. After this time the excess solution is removed (poured off or filtered off, but not washed off) and several drops of 1% hydrochloric acid are placed on the thin section or the grain for 7-8 seconds. The thin section (or grain) is then washed thoroughly. As a result of this procedure ankerite and calcite are stained dark blue, aragonite is unevenly stained blue, dolomite becomes pale blue, and siderite acquires a greenish tone.

Reaction with copper nitrate. A powder of carbonate mineral is boiled in a 5% solution of $Cu(NO_3)_2$ for 2-3 minutes (the reaction of the medium is slightly acid). The solution is then poured off and the sediment is washed to remove any excess reagent. By this process calcite and aragonite are stained bright green, magnesite pale blue, ankerite pale green or lettuce green, dolomite not at all or barely pale green; grains of siderite and breunnerite retain their individual colors.

Reaction with methylene blue. A test tube containing fine powder of a carbonate mineral (≤ 0.1 mm) and 2-3 cm^3 of distilled water is shaken and 2-3 cm^3 of 0.001% solution of methylene blue is added. If the carbonate is ankerite, the solution becomes light blue, the sediment lilac colored, and there is no change on adding a saturated solution of KCl; magnesite acts similarly under these conditions, but finely powdered aragonite gives a light blue solution and dark purple sediment; the other carbonate minerals produce no colored solutions in this procedure, or they make the solution a turbid white.

Ankerite is formed in sedimentary deposits chiefly during diagenesis of sediments (it is also widely developed by hydrothermal processes). In particular, ankerite is widespread in concretions and in disseminated grains in oil-bearing (probably oil-producing) and coal-bearing strata. It is found in sedimentary ankeritic ores (beds and seams).

Magnesite — $MgCO_3$. An anhydrous carbonate of magnesium, containing 47.6% MgO and 52.4% CO_2, and crystallizing in the trigonal system. The hardness ranges from 3.5-4.0 to 4.5, and the specific gravity is 2.9-3.12. It is white, yellowish or grayish white, light gray, sometimes snow white or, contrastingly, brownish. The indices of refraction are No = 1.700-1.717 and Ne = 1.507-1.515; No—Ne = 0.193-0.202; the mineral is uniaxial, optically negative, and has distinct pseudoabsorption. Isomorphous admixtures of $FeCO_3$ increase the refractive indices and the birefringence (the beginning of the isomorphous series $MgCO_3$—$FeCO_3$). At an $FeCO_3$ content of 5 to 30% the mineral is called breunnerite.

Feigl and Leitmeier [1928] suggested a reaction with diphenyl carbazide (in an alkaline environment) to distinguish magnesite (or hydromagnesite) from dolomite in powder form. M. I. Faneev [1936] proposed a technique of using the indicated reaction on thin sections, where the structural relations of the minerals could be observed.

G. L. Piotrovskii [1934] suggested using a reaction with paranitroazobenzene resorcinol to stain magnesite in thin sections: 25 mg of this reagent is dissolved in 500 cm^3 of water and 500 cm^3 of alcohol and caustic alkali is added until the solution becomes dark blue. A thin section of carbonate rock is submerged in this solution for 3-5 mintues. During this time magnesite is stained dark blue; other carbonates do not change color. The blue color is unstable and disappears rather quickly; the thin section should therefore be examined under the microscope immediately after staining.

Magnesite does not dissolve in cold acid, but it will dissolve when boiled in 30% of concentrated hydrochloric acid.

Sedimentary magnesite (hydrothermal magnesite is also known) is found in ancient lagoonal or salt-lake deposits, in the zone of surface weathering on masses of ultrabasic rocks, in disseminated grains in sequences of rock salt and easily soluble salts, and also in some anhydrite rocks.

In an ancient weathering zone on serpentinites in the Southern Urals [Ginzburg and Rukavishnikova, 1951] three varieties of magnesite have been distinguished: a) crystalline, 2) "amorphous," (more precisely, dense, cryptocrystalline, generally colloform), and c) colored, opalized, representing opalized colloform magnesite.

The crystalline magnesite in the Southern Urals shows the following range of refractive indices: No = 1.700-1.726 and Ne = 1.509-1.527; No—Ne = 0.191-0.199. This material is hydrothermal in origin.

The cryptocrystalline magnesite has been formed both by hydrothermal processes and by surface weathering. Typical colloform cryptocrystalline varieties form a third generation, clearly supergene.

The indices of refraction of the cryptocrystalline magnesite in the Southern Urals are No = 1.700-1.713 and Ne = 1.509-1.510; No−Ne = 0.190-0.204. The opalized magnesite is not a mineral, but a mixture of colloform magnesite and opal. All the samples of magnesite from the weathering zone in the Southern Urals show traces of hydromagnesite on chemical analysis.

Siderite —$FeCO_3$. A carbonate mineral of the trigonal system, consisting of 62.1% FeO (i.e., 48.3% Fe) and 37.9% CO_2, also called iron-spar. Its hardness is 3.5-4.5, generally 4.0, and the specific gravity is 3.8-3.9. Crystals are rhombohedral or of distorted rhombohedral aspect. The color is yellowish white, grayish, gray, blueish gray, greenish gray, dark gray, brownish gray, brown, and yellow; the brown and yellow tones are due to oxidation during weathering. The refractive indices are No = 1.872-1.875 and Ne = 1.633-1.634; No−Ne = 0.232-0.242; the mineral is uniaxial, negative, and has marked pseudoabsorption emphasizing the normal light brown tones or the weak colors of the grains in thin section.

There are known to be isomorphous series of minerals of siderite—magnesite and siderite—rhodochrosite; it is ordinarily believed that the siderite—calcite series is broken in the middle. If the content of $MgCO_3$ in siderite is more than 5% and up to 30%, i.e., a $FeCO_3$ content of 95 to 70%, the mineral is called sideroplesite; if the $MnCO_3$ is more than 5% and less than 30%, the mineral is called manganosiderite. And calcareous siderite, with a content of isomorphous $CaCO_3$ between 5 and 20% (sp. gr. up to 3.41), is called siderodot. The other of this interrupted series is represented by ferrocalcite, containing from 1 to 10-15% $FeCO_3$.

Siderite is found in sedimentary rocks in the following varieties: as microgranular (Fig. 8) or fine-grained rhombohedral (Fig. 9) matrix of rocks and as part (or one of the components) of cement in siltstones and, rarely, in sandstones; in disseminated rhombohedral or rounded-rhombohedral grains in argillaceous rocks; as prismatic grains; frequently in microspherulites (occasionally multilayered), radiolites (Fig. 10), and similar formations.

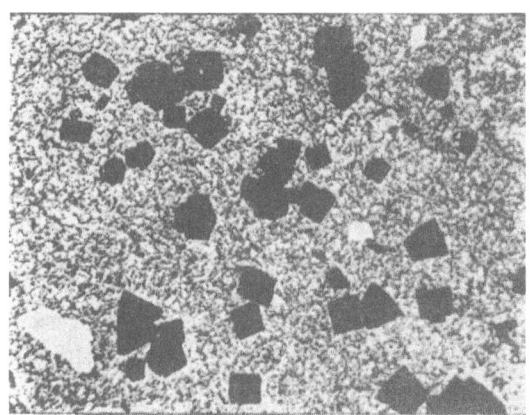

Fig. 8. Microgranular siderite (the gray background) with small crystals of pyrite (black inclusions). × 35. (From Cayeux, 1916.)

In addition to the pseudoabsorption and the normal brown tones in thin section, siderite is characterized also by lack of twinning and by high relief, the values of all indices being distinctly higher than that for Canada balsam; this latter property easily permits one to distinguish siderite in thin section from all other ordinary sedimentary carbonates. The mineral dissolves in hot 10% hydrochloric acid. A film of Turnbul blue (ankerite gives a dark blue film) is formed on the surface of a piece or grain of siderite moistened with 1% solution of potassium ferricyanide, $K_3Fe(CH)_6$, that has been acidified with several drops of strong hydrochloric acid. According to N. V. Logvinenko and N. K. Zabolotnaya, (1954), siderite is not stained in this section by any of the organic dyes nor by most of the inorganic reagents (either it does not change or it becomes somewhat darker, in some reactions acquiring a greenish tone). The mineral shows a characteristic thermal curve with an endothermic effect due to dissociation of $FeCO_3$ at 475-540° and a subsequent exothermic effect (600-890°) corresponding to the oxidation of FeO.

Siderite is chiefly a syngenetic mineral (the period of diagenesis of sediments) and is widespread in sedimentary rocks; it forms individual layers, lenses, and concretions, is found in disseminated grains in argillaceous rocks, or forms part of the cement in siltstones. Argillaceous and marly spherical concretions of siderite are called spherosiderites. Frequently several generations of diagenetic siderite may be recognized (2 or 3); and epigenetic siderite is also found.

Siderite of sedimentary and metamorphosed sedimentary, as well as hydrothermal, origin locally forms economic deposits. It forms valuable iron ore if the content of harmful impurities is low (sulfur, phosphorus, etc.).

Siderite forms in weakly reducing (to neutral) and reducing environments. It is therefore no accident that disseminated siderite is found with disseminated FeS_2 in characteristic occurrence in some sand-silt-clay littoral or shallow-water strata that contain abundant organic material. Siderite, like other iron minerals occurring in nature, is an indicator of the oxidizing-reducing conditions.

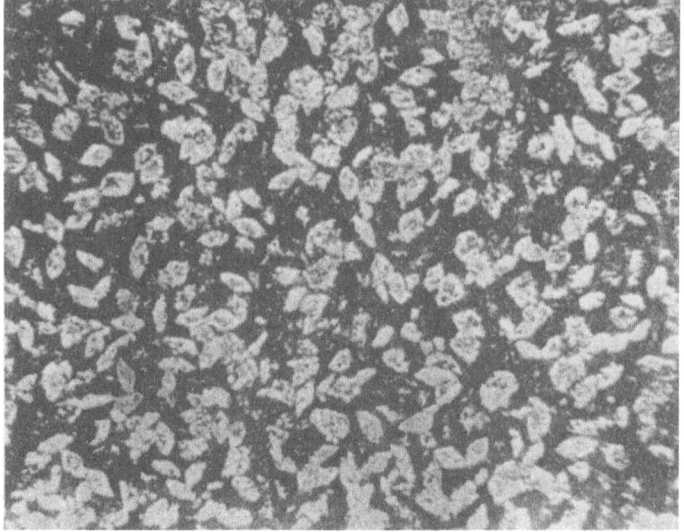

Fig. 9. Very fine-grained rhombohedral siderite, ×100. (From Cayeux, 1916.)

Siderite — Magnesite Isomorphous Series

The varieties distinguished in the siderite—magnesite isomorphous series are shown in Table 6.

Recent studies of the minerals in sedimentary rocks have shown that authigenic sideroplesite and pistomesite are frequently present, but that they were previously identified as "siderite"; breunnerite has also been found in sedimentary deposits.

Let us examine breunnerite briefly. It is a magnesium-iron carbonate mineral in the trigonal system. Its hardness is 4.0-4.5, and the specific gravity ranges from 3.0 to 3.2. The color is white, grayish white, yellowish white, and yellowish. The indices of refraction increase with increase in the content of isomorphous $FeCO_3$: at 9% $FeCO_3$ No = 1.707 and Ne = 1.517; at 30% $FeCO_3$ No = 1.725 and No—Ne = 0.190-0.192. The mineral is uniaxial and negative. Powdered breunnerite dissolves when boiled in 30% hydrochloric acid.

Fig. 10. Radiolitic siderite. × 35, nicols crossed. (From Cayeux, 1916.)

Data on staining the powders of carbonate minerals in this series, by means of a reaction with eosin (on boiling), have been given by N. V. Logvinenko and N. K. Zabolotnaya [1954].

According to these authors, breunnerite is best identified by a reaction with eosin (tetrabromofluorescein), $Ca_{20}H_8O_5Br_4$, in an alkaline environment. The reagent is prepared in the following way: a test tube is filled half full of alcohol, and 1-2 g eosin is dissolved in it while being heated; about 3 ml 25% potassium hydroxide is then added. The powdered mineral is immersed in the prepared solution and boiled for two minutes. After this the solution is poured off and the reagent is washed from the sediment until the water is no longer discolored.

As a result of this reaction breunnerite is stained a pale rose color, magnesite a bright rose color, siderite becomes somewhat darkened, and the other carbonates (including ankerite) are not stained.

Reaction with paranitroazobenzene resorcinol, $NO_2C_6H_6N \cdot NC_6H_3(OH)_2$, in an alkaline environment. Take a 25 mg/liter mixture of alcohol and water and introduce an alkaline excess (KOH or NaOH) to obtain a dark blue solution. The powdered carbonate mineral is boiled in this solution for 2-3 minutes. The solution is then poured off and the sediment is washed until the water is no longer discolored. As a result of this reaction breunnerite and dolomite acquire a dark blue stain, magnesite a bright blue and blue-green stain; ankerite, siderite, and the other carbonate minerals are not stained.

TABLE 6

Minerals in the Siderite—Magnesite Isomorphous Series

Mineral	$FeCO_3$ content, in %	$MgCO_3$ content in %	Sp. gr.	Refractive indices
Siderite	100—95	0—5	3.8—3.9	No=1.875, Ne=1.633
Sideroplesite	95—70	5—30	3.6—3.7	No=1.847
Pistomesite	70—50	30—50	3.3—3.5	No=1.804
Mesitite	50—30	50—70	3.1—3.3	No=1.769
Breunnerite	30—5	70—95	3.0—3.2	No=1.725
Magnesite	5—0	95—100	2.9—3.12	No=1.700, Ne=1.507

Reaction with Fe_2Cl_6 and $(NH_4)_2S$ in a weak acid environment is ineffective in identifying breunnerite. A thin section, polished section, or grain is treated with 10-12% solution of Fe_2Cl_6 for 20-25 seconds. After being washed, the sample is treated with $(NH_4)_2S$ for several seconds. The sample is then washed thoroughly. As a result of this reaction (Lemberg) calcite and aragonite give a black precipitate, dolomite in fine powder becomes dark green, breunnerite and magnesite acquire a greenish tone, and ankerite and siderite become slightly greenish.

Breunnerite has been recognized in sedimentary rocks only recently, in coal-bearing strata in concretions and in disseminated grains; it is commonly found with ankerite.

Rhodochrosite — $MnCO_3$. A mineral of the trigonal system, consisting of 61.7% MnO (47.8% Mn) and 38.3% CO_2 and also called manganese-spar; it commonly contains isomorphous admixtures of $CaCO_3$, $MgCO_3$, or $FeCO_3$. The hardness is 3.5-4.5, and the specific gravity is 3.6-3.7. It is generally rose colored, rarely raspberry colored; the color becomes paler with increase in the isomorphous admixture of $CaCO_3$; fine-grained and earthy varieties are white with a very slight rosy cast; when exposed to air the mineral loses the rose color and becomes brown in time, because of the oxidation of the $MnCO_3$; on heating the mineral becomes black. The indices of refraction are No = 1.817 and Ne = 1.597; No−Ne = 0.220. Rhodochrosite is uniaxial and optically negative.

Apart from the color, rhodochrosite is distinguished from calcite by a test for manganese and by the fact that both indices of refraction for this mineral are higher than that for Canada balsam. These indices become somewhat lower with isomorphous admixtures of $FeCO_3$ and markedly lower with $CaCO_3$. On heating, rhodochrosite dissociates in the interval 500-600° (an endothermic effect appears on the thermal curve) and an exothermic reaction occurs between 600 and 900°, resulting from the oxidation of MnO. $MnCO_3$ dissolves in the bicarbonate form, but much more weakly than $FeCO_3$; rhodochrosite dissolves in hot 10% hydrochloric acid.

Rhodochrosite is generally found in sedimentary deposits in compact masses with finely crystalline radiating structure, commonly in nodular and botryoidal concentrically layered structures. In nature $MnCO_3$ forms continuous isomorphous series with $FeCO_3$ and, especially, with $CaCO_3$. Varieties of the latter series (calcium rhodochrosite and manganocalcite) commonly alternates with rhodochrosite in the sections of manganese sedimentary ore deposits.

Rhodochrosite — Calcite Isomorphous Series

In the rhodochrosite — calcite isomorphous series, the varieties are distinguished as shown in Table 7.

The thermal curves for calcium rhodochrosite show a first endothermic effect ending approximately at 600°; with increase in the isomorphous admixture of $CaCO_3$ this effect becomes less distinct and at a content of $CaCO_3$ greater than 30% a second weak endothermic effect is noted at 700-740°; i.e., the following exothermic effect is inhibited and restricted within a temperature interval that gives it a peak-like character. This aspect of the thermal curve is due to the dissociation of $CaCO_3$, which occurs simultaneously with the dissociation of $MnCO_3$ and increases at higher temperature.

Rhodochrosite, calcium rhodochrosite, and manganocalcite are widely distributed in sedimentary carbonate manganese ores; opal-rhodochrosite ores form farther from the shore of the basin than oxide manganese ores do.

TABLE 7

Minerals of the Rhodochrosite—Calcite Isomorphous Series

Mineral	$MnCO_3$ content, in %	$CaCO_3$ content, in %	Sp. gr.	Indices of refraction	Hardness
Rhodochrosite	100—95	0—5	3.6—3.7	No=1.817, Ne=1.597	3.5—4.5
Calcium rhodochrosite	95—50	5—50	3.3—3.5	No=1.772 (with 70% $MnCO_3$) No=1.738 (with 50% $MnCO_3$)	3.5—4.0
Manganocalcite	50—5	50—95	2.9—3.3	No=1.707 (with 30% $MnCO_3$)	3.5—4.0
Calcite	0—5	100—95	2.72	No=1.658, Ne=1.485	3.0

Siderite — Rhodochrosite Isomorphous Series

The subdivisions of the siderite—rhodochrosite isomorphous series are shown in Table 8.

In Russian literature oligonite has been referred to the widest range (approximately 75-25% $FeCO_3$ and 25-75% $MnCO_3$); i.e., it corresponds approximately to oligonite and manganosphaerite according to A. N. Winchell and H. Winchell.

Sedimentary oligonite, crystallizing in the trigonal system is most frequently mentioned in the literature. Its hardness is 3.5-4.5, and the specific gravity is 3.7-3.8. The color is white, yellowish gray, or rose. With increase in the iron content the refractive indices become somewhat higher: at a Mn:Fe ratio of 4:1, No=1.828 and Ne = 1.606, and No—Ne = 0.222; when Mn:Fe = 1:5, No = 1.849, Ne = 1.615, and No—Ne = 0.234; the mineral is uniaxial and optically negative.

E. Larsen and H. Berman [1937] have shown that oligonite with a composition of 35.28% MnO, 26.18% FeO, and 37.98% CO_2, i.e., consisting of 57% $MnCO_3$ and 42% $FeCO_3$, has the refractive indices No = 1.840 and Ne = 1.695.

Oligonite is a common mineral in carbonate ores in sedimentary manganese deposits along the eastern slope of the Northern Urals; it also occurs in hydrothermal deposits of lead-zinc sulfide ores. Ferrorhodochrosite and manganosiderite, according to A. G. Betekhtin, are little-used synonyms of oligonite.

TABLE 8

Minerals of the Siderite—Rhodochrosite Isomorphous Series (from A. N. Winchell and H. Winchell, 1953).

Mineral	$FeCO_3$ content, in %	$MnCO_3$ content, in %
Siderite	100—95	0—5
Manganosiderite	95—70	5—30
Manganospaerite	70—50	30—50
Oligonite	50—30	50—70
Ponite	30—5	70—95
Rhodochrosite	5—0	95—100

Aragonite Subgroup

Aragonite—$CaCO_3$. This mineral has the same chemical composition as calcite, but it frequently contains a marked admixture of strontium (strontioaragonite or mossotite) or other elements. It is in the rhombohedral system; apart from the crystal system, it is distinguished from calcite by greater specific gravity (2.93-2.95) and hardness (3.5-4.0). The crystals are prismatic or acicular. The mineral is white, yellowish white, or colorless. Cleavage is practically absent. Typical aragonite has the following indices of refraction: Ng = 1.686, Nm = 1.681, and Np = 1.530; Ng—Np = 0.156; it is biaxial, negative, and has negative elongation; prismatic grains show parallel extinction; pseudoabsorption is marked.

Aragonite is also distinguished from calcite by its greater solubility in water and by its somewhat poorer solubility in hydrochloric acid. Under the microscope it is distinguished from calcite only in large grains, when it is possible to determine the biaxial nature of the mineral. Meigen's reaction, generally used to distinguish aragonite from calcite, is not sufficiently reliable. It involves treating the powdered mineral with a 10% solution of $Co(NO_3)_2 \cdot 6H_2O$, boiling the mixture for several minutes (aragonite should be stained a purple color, and calcite should remain colorless on boiling up to 10 minutes). A reaction proposed by S. Tugutt [1910] deserves attention: powdered mineral is treated for one second with decinormal (1.7%) solution of $AgNO_3$; the powder is then washed with water, treated with a 20% solution of K_2CrO_4, and again washed. As a result of this test coarse powdered fragments of aragonite acquire a distinct red color. Calcite is stained the same color if it is left in the decinormal solution of $AgNO_3$ for half an hour [Tatarskii, 1955].

Aragonite is found in sedimentary rocks in microgranular, rarely fibrous aggregates (some shells and oolites). The shells of modern gastropods, pteropods, cephalopods, and some pelecypods consist of aragonite, which is unstable in the surficial zones of the sedimentary rock layer and gradually changes to calcite; the same statement applies to remains of hydroids and oolites.

This process of alteration also explains the recrystallization of the skeletal parts of Paleozoic representatives of the above-listed groups of organism. The recrystallization has been the reason for the concept that aragonite is rare in ancient rocks.

Ktypeite is similar to aragonite.

Ktypeite. This mineral form was identified by Lacroix in recent calcareous oolites. It has a birefringence less than 0.040-0.050 and is oriented in the oolites with Ng parallel to a radium and Np tangent to the concentric layers; the chemical composition is the same as that of aragonite. Sorby [1879] and most recent authors [Zavaritskii, 1929; Zavaritskii and Mikheev, 1948] here considered ktypeite to be a cryptocrystalline variety of aragonite in present-day oolites, the indivisible fibers of which are arranged tangential to the surface. When indivisible crystals aragonite are arranged in a single plane tangent to the surface of growth in an oolite, we should observe an interference color between Ng−Np and Nm−Np or Ng−Nm of aragonite, or nearly so if the crystals are not oriented strictly in a single plane.

Vaterite − $\mu CaCO_3$. This form was first obtained by Vater artificially. It forms fibrous, hexagonal-tabular, or lenticular carbonate crystals belonging to the hexagonal system, being distinguished by its structure from calcite and aragonite; it apparently develops by crystallization of $CaCO_3$ gel. The indices of refraction are Ne = 1.640-1.650 and No = 1.550; the mineral is uniaxial and positive. Vaterite changes to calcite with time. Maier believed that vaterite and aragonite form the shells of several modern gastropods, vaterite being the initial crystalline phase and then changing to aragonite [Chukhrov, 1955]. Diagnostic properties of vaterite have not been sufficiently determined.

Strontianite − $SrCO_3$. This mineral belongs in the orthorhombic system. Its hardness is 3.5-4.0, and the specific gravity 3.6-3.8. It is colorless or has a greenish, yellowish, or grayish cast. The refractive indices are Ng = 1.666-1.668, Nm = 1.664-1.667, Np = 1.516-1.520; Ng−Np = 0.148-0.150; the mineral is biaxial and optically negative. It is distinguished from aragonite by greater specific gravity and by the red color in a flame. Strontianite is frequently found in hydrothermal deposits (together with celestite, barite, calcite, and sulfides) and in sedimentary rocks (limestones, dolomites, and marls); it occurs as an authigenic mineral in geodes and veinlets. The mineral may be replaced by celestite.

Witherite − $BaCO_3$. The mineral is in the orthorhombic system. Its hardness is 3.0-3.5, its specific gravity 4.2-4.3. It is colorless, white, gray, or yellowish. The indices of refraction are Ng = 1.677, Nm = 1.676, and Np = 1.529; Ng−Np = 0.148; the mineral is biaxial and negative. It is distinguished from aragonite and strontianite by greater specific gravity and by a yellow-green color in the flame test. It is common in hydrothermal deposits, and is also known to be of exogene origin. It occurs in crystals, in spherical and nodular masses, and in veins and fibrous and lamellar aggregates.

Cerussite − $PbCO_3$. A mineral in the orthorhombic system. The hardness is 3.0-3.5; the specific gravity 6.4-6.6. It is white with grayish, yellowish, or brownish tones. The refractive indices are Ng = 2.078, Nm = 2.076, and Np = 1.804; Ng−Np = 0.274; the mineral is biaxial, optically negative, and has an optic angle of 8°. It is distinguished from the other carbonates by greater specific gravity and by higher indices of refraction. It is found chiefly in the oxidized zones of lead-zinc sulfide deposits, commonly with galena and anglesite ($PbSO_4$). It may also be a low-temperature hydrothermal mineral.

Cerussite generally occurs in compact granular masses, but also in incrusting, cryptocrystalline, and earthy forms; fibrous varieties are occasionally encountered.

Basic Carbonates

Azurite and malachite belong in this subgroup of sedimentary minerals.

Malachite — $Cu_2(CO_3)(OH)_2$, or $CuCO_3 \cdot Cu(OH)_2$. This mineral crystallizes in the monoclinic system, but it generally occurs in nodular-incrusting formations with radiating-fibrous, rarely concentrically zoned structures; earthy varieties are also known ("copper earth"). The hardness is 3.5-4.0, the specific gravity 3.9-4.1. The mineral is characteristically a bright green. It is transparent in thin section. The indices of refraction are Ng = 1.909, Nm = 1.875, and Np = 1.655; Ng−Np = 0.254; the mineral is biaxial, negative, and the extinction angle (c$^\wedge$Np) is 23°.

Malachite is a characteristic mineral in the zone of oxidation of copper sulfide deposits, especially when such deposits occur in limestones or when the primary ore contained large quantities of carbonates. The mineral forms by replacement of carbonates, by filling cavities, and may be found as pseudomorphs after azurite and cuprite.

Malachite is widespread in typical sedimentary cupriferous sandstones, such as in Tataria [Miropol'skii, 1938] and in the Donbas.

Azurite — $Cu_3(CO_3)_2(OH)_2$, or $2CuCO_3 \cdot Cu(OH)_2$. A mineral in the monoclinic system. The hardness is 3.5-4.0, the specific gravity 3.7-3.9. The color is dark blue and azure blue; in earthy forms it is light blue. The indices of refraction are Ng = 1.838, Nm = 1.758, and Np = 1.730; Ng−Np = 0.108; the mineral is optically positive and the extinction is inclined, c Ng = 13°. Azurite is always found in small quantities in association with malachite. Apparently it forms in somewhat less alkaline environments than malachite. It occurs in typical sedimentary cupriferous sandstones, as in Tataria [Miropol'skii, 1938] and in the Donbas.

Hydrous Carbonates and Basic Hydrous Carbonates

Of the sedimentary hydrous carbonates, we have described here only natron and nesquehonite, and of the basic hydrous carbonates, only hydromagnesite.

Natron — $Na_2CO_3 \cdot 10H_2O$. A decahydrate sodium carbonate, crystallizing in the monoclinic system. Its hardness is 1.0-1.5, the specific gravity 1.4-1.47. Crystals have a rhomboidal tabular form. The mineral is colorless, white, or gray. The indices of refraction are Ng = 1.440, Nm = 1.425, and Np = 1.405; Ng−Np = 0.035; the mineral is biaxial and negative. Natron is generally found in granular masses. Its precipitates from saturated solutions of Na_2CO_3 at atmospheric pressures within the temperature range from − 2 to + 32°; it quickly loses water in air and changes to thermonatrite, $Na_2CO_3 \cdot H_2O$. It dissolves easily in water.

Natron is a typical sedimentary or exogene mineral. It accumulates in large masses from sulfate-bicarbonate waters in some salt lakes that are rich in sodium.

Natron forms in small quantities on the surfaces of unconsolidated rocks and soils (efflorescenses and incrustations) in hot, arid climates.

Nesquehonite — $MgCO_3 - 3H_2O$. A trihydrate magnesium carbonate* that crystallizes in the orthorhombic system. The hardness is 2.5, the specific gravity 1.83-1.85. The mineral is white or colorless. Crystals form long prisms. The indices of refraction are Ng = 1.526, Nm = 1.501, and Np = 1.412; Ng−Np = 0.114; the mineral is biaxial and negative. It occurs in prismatic crystals in radiating (acicular) aggregates. It forms by dehydration

---

* According to Davis and D'Ans, nesquehonite is more properly considered to be a "basic bicarbonate of magnesium" with the formula $Mg[(OH)(HCO_3)]_2 \cdot 2H_2O$ or $Mg(OH)_2 \cdot Mg(HCO_3)_2 \cdot 4H_2O$; in the indicated treatment it is a hydrous basic magnesium carbonate (as indicated by the nature of the thermal curve).

of $MgCO_3 \cdot 5H_2O$, i.e., of lansfordite, and in air it slowly changes to magnesite. It gives a characteristic thermal curve: a) below $220-225°$ an endothermic effect due to separation of water of crystallization; b) near $410-440°$ a very small endothermic effect, apparently due to decomposition of the hydroxyl group; c) from 480 to $508°$ an endothermic depression in the curve indicating the beginning of CO_2 separation; d) at $508-525°$ an exothermic "explosive" effect, probably corresponding to the crystallization of amorphous MgO that appeared during decomposition of brucite, the MgO becoming periclase in this interval; e) from 530 to $575°$ an endothermic effect, corresponding to the elimination of CO_2 from the main part of the magnesite structure.

Nesquehonite has been found in a number of coal mines. Recently it was discovered with hydromagnesite and the more widespread magnesite in Lower Permian lagoonal deposits in the Kuibyshev and Saratov trans-Volga region [Frolova, 1955]. Masses of magnesium carbonate generally consist of pelitomorphic magnesite and, in considerably less quantity, isolated prismatic crystals of nesquehonite and hydromagnesite. Nesquehonite is found in the indicated region in sulfate-carbonate rocks. Magnesite is associated either with anhydrite and dolomite or with minerals of the halite group in sites of chloride cementation.

Hydromagnesite — $4MgCO_3 \cdot Mg(OH)_2 \cdot 4H_2O$. A hydrous basic magnesium carbonate with the structural formula $Mg_5(CO_3)_4(OH)_2 \cdot 4H_2O$ and crystallizing in the orthorhombic system; according to some authors the mineral is monoclinic, pseudo-orthorhombic. The hardness is 3.5, the specific gravity 2.13-2.18. The mineral is white. It is found in compact (microgranular or fibrous) chalk-like masses, in powdery and tuff-like masses, and also in small crystals forming fan-shaped radiating aggregates or forming small spherulites.

The indices of refraction are Ng = 1.538-1.545, Nm = 1.527-1.531, and Np = 1.515-1.523 up to 1.527; Ng—Np = 0.008-0.022-0.03; the mineral is biaxial and positive.

In contrast to magnesite, which has a single endothermic peak on the thermal curve (at $610-620°$, on some at $650°$), hydromagnesite has two distinct effects, and three in all: the first from 270 to $380-450°$, the second (small) at $450°$, and the third beginning at $550°$ and having a maximum at $600°$. The first endothermic effect is due chiefly to elimination of water of crystallization ($4H_2O$), the second (small) effect to the separation of hydroxyl water in brucite, and the third to the final dissociation of carbonate; at $540°$ there occurs an exothermic reaction corresponding to the recrystallization of amorphous MgO into periclase, the MgO being derived from the decomposition of brucite.

Hydromagnesite forms chiefly by weathering of ultrabasic magnesian magmatic rocks, partly as a hydrothermal mineral.

Hydromagnesite in the weathering zone of serpentinites in the Southern Urals, according to chemical analyses, has the formula $3.5 MgCO_3 \cdot Mg(OH)_2 \cdot 4.5 H_2O$.

Hydromagnesite is found with cryptocrystalline and, in part, with typical colloform magnesite; magnesium carbonate is normally precipitated as the basic carbonate (hydromagnesite), being changed gradually to magnesite as it loses water.

3. SEDIMENTARY SILICATE GROUP

This group includes the following authigenic minerals in sedimentary rocks: iron silicates, aluminosilicates of the "clay group" (including hydrous magnesian silicates and aluminosilicates), hydrous aluminosilicates of calcium and sodium (zeolites and analcime), authigenic feldspars, and copper silicates.

Iron Silicates

These minerals are distinguished by green color and are subdivided into two subgroups: potassium-bearing (glauconite and protoglauconite) and non-potassium-bearing (various iron chlorites and, in part, iron-magnesium chlorites).

Potassium-Bearing Iron Silicates

Glauconite — $K_{<1}(Fe^{3+}, Fe^{2+}, Al, Mg)_{2-3} \cdot [Si_3(Si, Al)O_{10}][OH]_2 \cdot nH_2O$. This is a hydrous silicate, chiefly with ferric oxide, partly with ferrous oxide; SiO_2 ranges from 40.0-47.6 to 52.9-56.0%, Fe_2O_3 from 16 to 22% (up to 27.9% in highly ferruginous varieties, but in aluminous varieties from 5-6 to 11-18%), Al_2O_3 from 4.0-4.5 to 10 and up to 21% (in aluminous varieties), FeO from 0.8-1.5 to 3.0% (up to 8.6-9.6% in some highly ferruginous varieties), K_2O from 4.0-5.0 to 7.5-9.6%, Na_2O from 0 to 3.3%, MgO from 1.6-2.0 to 4.1-4.6%, and H_2O from 4.9-6.0 to 8-13.5%. The aluminous varieties of the glauconite group are frequently called skolite (they have been called protoglauconite by A. V. Kazakov); hydrothermal ferruginous varieties with a high MgO content (from 3.84-4.40 to 8.54-9.32%) have been called celadonite.

Various disputes concerning the independence of glauconite as a mineral species have recently been resolved by x-ray analysis, which has shown that the mineral has a crystal lattice similar to that for biotite; it has consequently been assigned by some authors to the hydromica group.

Glauconite is in the monoclinic (?) system. Its hardness is 2.0-3.0; its specific gravity ranges from 2.2-2.5 to 2.85-2.90. It is characteristically green or dark green. The index of refraction in ordinary grains with micro-aggregate polarization is near 1.59-1.61; in general N = 1.57-1.63, $Ng \cong Nm = 1.61-1.63$, Np = 1.59-1.60; $Ng-Np = 0.020-0.030$ (observed on cleavage plates of glauconite); the mineral is biaxial, negative. The refractive indices for skolite are Ng = 1.586; Nm = 1.582, Np = 1.559; $Ng-Np = 0.027$. For celadonite the indices are Ng = 1.638, Nm = 1.630, and Np = 1.608; $Ng-Np = 0.030$.

A. V. Kazakov and L. I. Gorbunova have found three principal types of glauconite and glauconite-like grains in the Jurassic and Lower Cretaceous rocks of the Russian platform [1947]:

a) dark green, in sandy shelf deposits (20% FeO_{23}; 6.7% K_2O; about 6% Al_2O_3; sp. gr. 2.7-2.9; N = 1.59);

b) yellow-green, from sediments at greater depth (17% Fe_2O_3; 4-5% K_2O; about 10% Al_2O_3; sp.gr. 2.6-2.8; N = 1.57);

c) slightly greenish-yellowish, colorless in thin section (5-8% Fe_2O_3; 16-18% Al_2O_3; 2-2.5% K_2O; sp.gr. 2.4-2.5; N = 1.54).

The glauconite-like grains of type "c" are not properly glauconite, but have been called protoglauconite, or alpha-glauconite, by A. V. Kazakov. According to this author the protoglauconite was deposited from land and at greater depth than the glauconite; it is distinguished from the latter by a smaller content of Fe_2O_3 and, at the same time, by a lower ratio of Fe_2O_3:FeO, a smaller content of potassium, a higher content of Al_2O_3, a lower specific gravity, and lower indices of refraction. However, V. A. Makhinin [1951] has concluded that in the Oligocene rocks of the Ukraine the aluminous variety of glauconite was deposited near shore and the ferruginous variety was formed at some distance from the shore line.

N. S. Shat'skii [1955] believes that it is not a greater depth of water that determines the various types of glauconite and protoglauconite, but that it is the nature of the enclosing rocks (sediments).

A great variety of forms of glauconite have been observed:

a) oval and irregularly lobed grains with microaggregate polarization are most widespread and characteristic (Fig. 11), i.e., microconcretions of numerous fine grains superimposed on each other within the thickness of a single thin section (the material has the character of a crystallized gel);

b) alteration to glauconite, along fractures and in various degrees, of grains of aluminosilicates that possess cleavage especially biotite;

c) filling foraminiferal chambers;

d) replacing skeletal remains of organisms (echinoids and others);

e) forming segregations or crusts on the surface and in cavities or fractures of fragmental grains and ordinary microaggregates of glauconite grains;

f) pigmenting, consisting of fine disseminated segregations without distinct outlines;

f) cement, of microaggregate or other types.

During weathering, glauconite grains become brown, changing into iron hydroxides and ferruginous clay.

In thin section, using only the polarizer, one may observe various internal structures in glauconite grains; the following are distinguished: a) homogeneous, b) granular, c) globular, d) parted along cleavage traces, e) rims of glauconite, normally fibrous or with traces of radiating cleavage and zonal structure.

With nicols crossed the internal structures of glauconite grains also exhibit differences: a) grains with microaggregate polarization are most abundant; b) one may rarely find grains with fibrous and web-like aggregation, showing aggregate extinction when the entire group goes extinct simultaneously; c) also, rarely, one may find radiating fibrous structure or normal fibrous structure.

It should be noted that glauconitic and argillaceous-glauconitic rocks are frequently found to consist of rounded grains of glauconite cemented by glauconite. Such glauconitic rocks were described by us from the Upper Cretaceous deposits in the vicinity of the Blyava station [1935₂], from the Khoper strata [1939₂], and from other horizons.

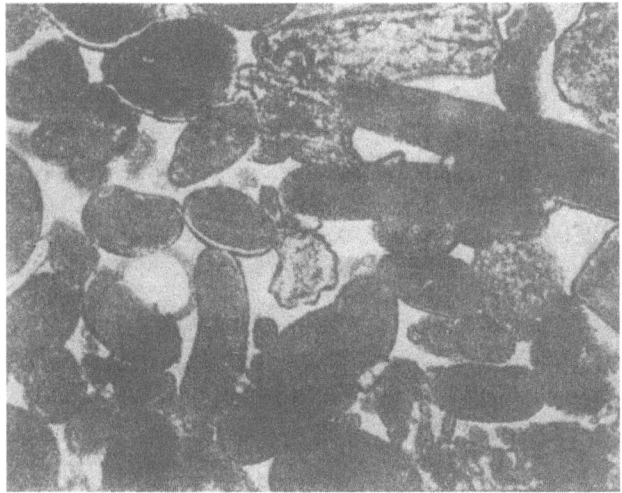

Fig. 11. Oval grains of glauconite with microaggregate polarization. × 60, nicols crossed. (From Hadding, 1932).

Glauconite is recognized by a number of characteristic features: by the green or dark green color, by the common "lobate" form of the grains, and by the general microaggregate extinction. It is distinguished from iron chlorites by the considerable content of potassium and, generally, by lower indices of refraction. The mineral decomposes in strong hydrochloric acid, leaving a very porous, white, siliceous framework, which preserves the general form of the grain. It does not dissolve in concentrated sulfuric acid, but it does break down in dilute (50%) sulfuric acid. The thermal curve of glauconite ordinarily shows three endothermic effects: a) the first effect (at 100-190°) results from the elimination of adsorbed water; b) the second (at 450-600°) is due to the loss of constitutional water; c) the third (at 775-975°) apparently reflects destruction of the crystal lattice and subsequent re-crystallization.

Adsorbed water is found in glauconite between structural layers of the crystal lattice, which are not so densely packed with cations as in ordinary micas (the mineral has a zeolitic character); this property determines the properties of absorption and ion exchange in glauconite.

Until recently glauconite was normally considered to be a characteristic mineral in marine sediments, it being understood that it might be reworked and might appear in typical continental (including alluvial) and lagoonal deposits.

At present, some authors [Lazarenko, 1956] believe that glauconite may be a product of weathering of various rocks and may even be produced hydrothermally, as well as being of marine sedimentary origin. From this viewpoint, the following distinctions may be made in the glauconite group:

1) ferruginous glauconites (typical glauconites), consisting of aluminum and iron silicates;

2) aluminous glauconites, consisting of aluminosilicates poor in iron (skolite, for example, in which $Al_2O_3 : Fe_2O_3 = 4 : 1$).

3) magnesian, strongly ferruginous glauconite, exemplified by celadonite (a hydrothermal mineral in amygdaloidal basalts).

It seems to us that both aluminous and magnesian-ferruginous glauconites belong to the glauconite group only in the broadest sense, being clearly distinguished from glauconite proper.

It is necessary to consider examples that are cited as proof that normal glauconite is formed under continental conditions.

It is most essential to refer to the statements that glauconite, in a later generation, may constitute part of the cement. But this by no means proves that glauconite cement is formed under continental conditions, especially since marine glauconitic rocks, consisting of rounded grains of glauconite with glauconite cement, are well known.

35

A second example is found in veinlets of aluminous glauconite (skolite) that occur in sandstones, and that may develop in the later stages of diagenesis, during katagenesis, and during epigenesis.

A third example, neoglauconite of A. V. Kazakov, is distinguished by a pale green color and represents, according to Kazakov, the weathering product of glauconite-phosphate cores, formed under the influence of atmospheric waters [Kazakov, 1938]; a similar occurrence has been cited by L. N. Formozova [1949]. Without considering the question of precisely when the neoglauconite was formed, it may be pointed out merely that the mineral is distinguished from typical glauconite not only by color but also by chemical composition. Neoglauconite, in the two indicated examples, in undoubtedly secondary; more particularly, it develops by fresh growth from the weathering products of glauconite.

A fourth example is the discovery of glauconite by M. G. Dyadchenko and A. Ya. Khatuntseva in Recent alluvium of the Irsha River (the Dnepr basin) and in the upper part of the eluvium on the Korosten intrusive mass. This example should be considered in more detail.

According to M. G. Dyadchenko and A. Ya. Khatuntseva [1956], it has been discovered that glauconite formed in the upper part of eluvial deposits on the flood plain of the Irsha River. A high content of glauconite was noted in eluvium on igneous rocks where this eluvium is covered directly by alluvial muds (loams, argillaceous silts, sandy loams). The color, indices of refraction, chemical composition, and x-ray photographs show that glauconite is present, having formed, in the opinion of the indicated authors, under continental conditions. The mineral contains 46.90% SiO_2, 21.45% Fe_2O_3, 3.16% FeO, 7.25% K_2O, 6.20% Al_2O_3, 4.20% MgO, and 2.60% hydroscopic H_2O; the loss during heating amounts to 6.32%. The specific gravity is 2.9; Nm = 1.629-1.619 and less. M. G. Dyadchenko and A. Ya. Khatuntseva believe that the glauconite was formed in this environment by replacement of potassium feldspars, plagioclase, apatite, and other minerals; most of the glauconite grains have typical rounded and nodular forms.

According to L. I. Karyakin and N. V. Logvinenko [1956], glauconite is generally an allogenic mineral in Ukrainian rivers, and its presence in the alluvium of the Irsha River is therefore natural. On the other hand, these authors have noted that the igneous rocks in the basin of the Irsha River were covered by marine waters during the Cretaceous and the Tertiary, as attested by the discovery of granitic gravel cemented by sandy-glauconitic material containing pelecypod shells. The replacement of potassium feldspars and other silicate minerals by glauconite may have occurred on the sea floor by halmyrolysis.

Thus, glauconite is a characteristic sedimentary mineral, forming chiefly in marine environments, but occasionally (neoglauconite) under continental conditions by weathering of sedimentary rocks containing abundant glauconite.

Glauconite is formed in the sea by different means: it precipitates from sea water as coagulated gel; it replaces aluminosilicates and several other minerals; it grows within foraminiferal chambers, it replaces $CaCO_3$ in the skeletal parts of organisms, and so forth.

Glauconite may thus form during sedimentation as well as during diagenesis of marine sediments. We shall not consider hydrothermal minerals of the glauconite group (in the broad sense) (see Lazerenko, 1954, 1956).

Marine glauconite is formed chiefly by precipitation from colloidal and molecular solutions, less commonly by replacement during halmyrolysis (submarine weathering) of fragmental minerals, and, in part, by biochemical means. The mineral is the product of a special marine mineralogical-geochemical facies, namely the glauconite facies, characterized by repeated microfluctuations in the oxidation-reduction boundary, i.e., characterized by repeated microfluctuations in the oxidation-reduction boundary, i.e., characterized by a struggle between oxidizing and reducing conditions, oxidizing conditions generally dominating [Pustovalov, 1933, 1940; Teodorovich, 1947; 1949, 1951, 1956]. N. S. Shat-skii [1955] has shown that marine glauconite is found in a variety of lithologic associations and complexes of very different ages, but it occurs chiefly in phosphorite-bearing formations.

Glauconite is a mineral of the shelf zone and the upper part of the continental slope; it forms where the shore is composed of magmatic rocks and where river mouths are not too near. Glauconite forms where bottom currents are strong, where sedimentation is retarded and even temporarily "reversed." Transgressions and regressions also favor the formation of glauconite, leading to movements of marine waters that constantly disturb the equilibrium. The temperature of formation of glauconite cannot be very low, since the mineral is abundant not only in siliceous and clastic complexes but also in calcareous sediments; in any case, the temperature of the bottom waters is decisive, not the temperature of the surface waters of the basin.

It should be stated that the optimum depth and temperature of water for the formation of glauconite in the seas and oceans have not been uniform throughout the various geologic periods (they have depended on the salinity of the marine waters, on the mineralization, and on other factors).

As N. S. Shat-skii pointed out [1955], glauconite-bearing deposits have been formed on platforms and, exceptionally, in geosynclines as well.

Non-Potassium Iron Silicates

These are green or dark green iron silicates but, in contrast to glauconite, contain no potassium, commonly being found in oolites or in aggregates with lamellar structure. The very widespread iron chlorites in sedimentary rocks are easily dissolved in hot hydrochloric acid, whereas the relatively rare iron-magnesium chlorites are more stable in an acid environment.

According to Tschermak two groups are distinguished among the chlorites: orthochlorites (true chlorites) and leptochlorites (pseudochlorites). Orthochlorites are understood to be isomorphous mixtures of two silicates having the composition of amesite ($H_4Mg_2Al_2SiO_9$) and serpentine ($H_4Mg_3Si_2O_9$), and being distinguished from leptochlorites by the considerably smaller content of iron. Recently several investigators have considered it more proper to assign all chlorites to one great series, leptochlorites being the ferruginous varieties corresponding to the orthochlorites; this view is based on x-ray data.

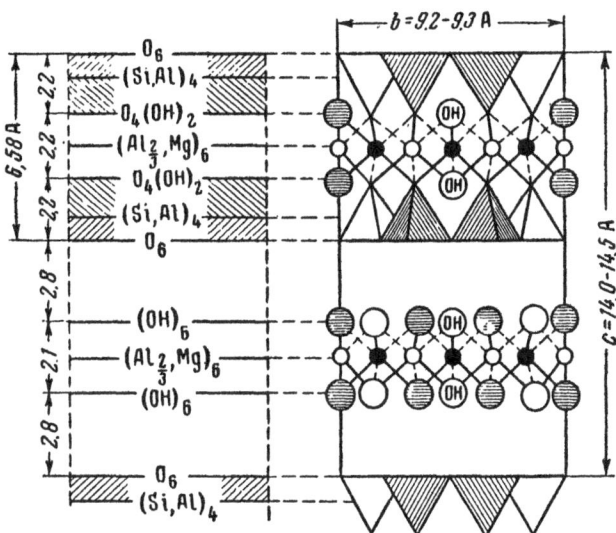

Fig. 12. The structure of chlorites. (From Pauling, 1930.)

The distribution of the layers in the lattice of chlorites is illustrated in Fig. 12. Two types of water are present in chlorites: the brucite type and the mica (or montmorillonite) type.

According to A. N. Winchell [1933, 1949], the chlorites are a group of minerals representing a mixture of antigorite (Ant), amesite (At), daphnite (Dn), and ferroantigorite (FeAnt). The structural formulas of the principal components of the chlorites are given below:

Antigorite (serpentine) . . . . $Mg_6(Si_4O_{10})(OH)_8$,

or $3MgO \cdot 2SiO_2 \cdot 2H_2O$;

Amesite . . . . $Mg_4Al_2(Al_2Si_2O_{10})(OH)_8$,

or $2MgO \cdot Al_2O_3 \cdot SiO_2 \cdot 2H_2O$;

Daphnite . . . . $Fe_4Al_2(Al_2Si_2O_{10})(OH)_8$, more accurately
$(Fe,Mg)_4Al_2(Al_2Si_2O_{10})(OH)_8$,

or $2(Fe,Mg)O \cdot Al_2O_3 \cdot SiO_2 \cdot 2H_2O$;

Ferroantigorite . . . . $Fe_6(Si_4O_{10})(OH)_8$, more accurately
$(Mg,Fe)_6(Si_4O_{10})(OH)_8$,

or $3(Mg,Fe)O \cdot 2SiO_2 \cdot 2H_2O$.

A. N. Winchell [1949, 1953] has proposed the following subdivision of the chlorites (Table 9).

TABLE 9

The Classification of Chlorites According to Chemical Composition (in %)
(From A. N. Winchell and H. Winchell)

Mineral	(3At + 3Dn) in %	(2Ant + 2Fe-Ant) in %	(2FeAnt + 3Dn) in %	(2Ant + 3At) in %
Antigorite	0—20	100—80	0—20	100—80
Jenkinsite*	0—20	100—80	20—40	80—60
Pennine	20—40	80—60	0—20	100—80
Delessite**	20—40	80—60	20—40	80—60
Clinochlore	40—60	60—40	0—20	100—80
Leuchtenbergite	40—60	60—40	20—40	80—60
Diabantite	40—60	60—40	40—60	60—40
Brunsvigite	40—60	60—40	60—80	20—40
Corundophilite	60—80	40—20	0—20	100—80
Prochlorite	60—80	40—20	20—40	80—60
Ripidolite***	60—80	40 20	40—60	60—40
Aphrosiderite***	60—80	40—20	60—80	40—20
Thuringite***	60—80	40—20	80—100	20—0
Amesite	80—100	20—0	0—20	100—80
Daphnite	80—100	20—0	80—100	20—0

*Iron antigorite according to some authors.

**Iron pennine according to some authors.

***Iron corundophilite according to some authors.

A. N. Winchell has proposed the following classification of the chlorites for petrographic studies (all members are in the monoclinic system) based on optical properties (Table 10).

The chemical and optical classification of chlorites and the properties of the minerals, according to A. N. Winchell, are shown in Fig. 13. The most common "iron-free" (more properly, very slightly iron-bearing) chlorites are pennine, clinochlore, and amesite; the slightly iron-bearing varieties are delessite and prochlorite; in A. N. Winchell's classification corundophilite is considered to be a slightly iron-bearing chlorite, but this view is not generally recognized.

In 1949, V. P. Ivanova proposed that the chlorites be divided into three groups: 1) magnesium, 2) iron-magnesium, and 3) iron (leptochlorites); some authors recognize a separate group of nickel chlorites.

V. P. Ivanova's group of magnesium chlorites is subdivided into two subgroups: 1) pennine-clinochlore, with Nm = 1.57-1.59, Ng−Np up to 0.011, an Orcel coefficient $S = (SiO_2)/(R_2O_3) = 2.5$-4.0, and two endothermic effects, the first being much greater than the second; 2) prochlorite-corundophilite, with Nm = 1.59-1.62, Ng−Np up to 0.011, S = 1.50-2.25, and two endothermic effects of equal intensity. According to V. P. Ivanova [1949], the chlorites of the iron-magnesium group are characterized by weak birefringence (0.002-0.007), Nm = 1.62-1.64, anomalous red interference color, strong pleochroism, vermicular structure, and a greater value for the first of the two endothermic effects. The chlorites of the iron group are most abundant in sedimentary rocks in which iron-magnesium chlorites are also present.

TABLE 10

Classification of the Chlorites According to Their Optical Properties (From
A. N. Winchell and H. Winchell)

Mineral	Optic sign	Nm	Ng − Np
Antigorite	−	1,55—1,58	0,004—0,010
Pennine	−	1,56—1,59	0,000—0,004
Jenkinsite	−	1,58—1,61	0,004—0,010
Delessite	−	1,59—1,61	0,000—0,004
Diabantite	−	1,61—1,63	0,000—0,004
Aphrosiderite	−	1,63—1,65	0,000—0,004
Daphnite	−	1,65—1,68	0,000—0,004
Brunsvigite	−	1,63—1,65	0,004—0,010
Thuringite	−	1,65—1,68	0,004—0,010
Pennine	+	1,57—1,61	0,000—0,004
Leuchtenbergite	+	1,59—1,61	0,000—0,004
Ripidolite	+	1,61—1,63	0,000 0,004
Clinochlore	+	1,57—1,59	0,004—0,010
Prochlorite	+	1,59 1,62	0,004—0,010
Amesite	+	1,58—1,61	0,010—0,015

Before discussing the crystallochemical classification of chlorites by D. P. Serdyuchenko (Table 11), let us call attention to the structural scheme of the chlorites according to Pauling [1930], which is found to agree with the composition and properties of natural chlorites (see Fig. 12).

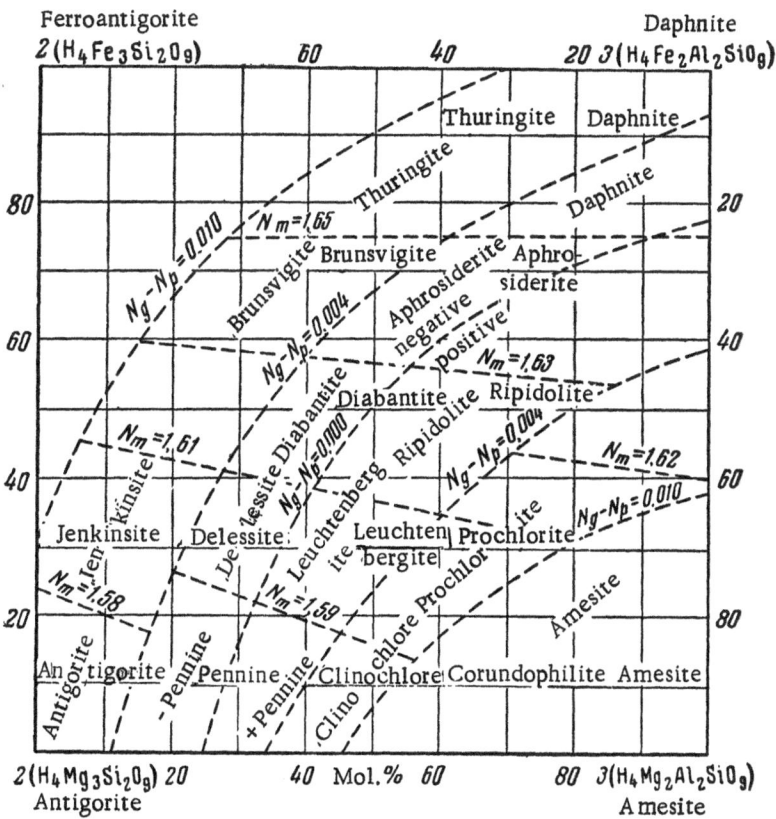

Fig. 13. The chemical and optical classification of minerals in the
chlorite group. (From A. N. Winchell, 1938, 1949.)

TABLE 11

Classification of the Chlorites According to D. P. Serdyuchenko [1953]

Crystallochemical criteria	Quantitative values	Nomenclature
Content of Si_{iv}	4.0-3.9 3.9-3.6	Serpentine Serpentine-chlorite } Serpentine
	3.6-3.4 3.4-3.2	Pennine Pennine-clinochlore } Pennine
	3.2-3.0 3.0-2.9	Clinochlore Clinochlore-prochlorite } Clinochlore
	2.9-2.7 2.7-2.6	Prochlorite Prochlorite-corundophilite } Prochlorite
	2.6-2.4 2.4-2.3	Corundophilite Corundophilite-metachlorite } Corundophilite
	2.3-2.1 2.1-2.0	Metachlorite Amesite } Metachlorite
With Si_{4-n}, a content of:		
Al_{iv}	n	Alumino-(chlorites)
$(Al, Fe^{\cdot\cdot})_{iv}$	n	Alumino-ferri-(chlorites)
$Fe^{\cdots}_{iv}$	n	Ferri-(chlorites)
$Y = R^{\cdot\cdot} + R^{\cdots}_{iv}$ (total cations with a coordination of six)	More than 6 Equal to 6 From 6 to 5.7 From 5.7 to 5.3 From 5.3 to 4.9 Less than 4.9	Serpentine, pennine, clinochlore, prochlorite, corundophilite, metachlorite } Aluminoserpentine, delessite, strigovite, chamosite, thuringite, aluminometachlorite
$a = Fe^{\cdots} : R^{\cdots}$	More than 0.6 From 0.6 to 0.3 Less than 0.3	Ferruginous Aluminous-ferruginous Aluminous (or without characteristic)
$c = Cr : R^{\cdots}$	More than 0.1 Less than 0.1	Chromian Chrome-bearing
$f = Fe^{\cdots} : R^{\cdot\cdot}$	More than 0.6 From 0.6 to 0.3 Less than 0.3	Ferruginous Magnesian-ferruginous Magnesian (or without characteristics)
$n = Ni : R^{\cdot\cdot}$	More than 0.1 Less than 0.1	Nickelian Nickel-bearing
$m = Mn^{\cdot\cdot} : R^{\cdot\cdot}$	More than 0.2 Less than 0.2	Manganian Manganese-bearing

D. P. Serdyuchenko has subdivided the chlorites into isomorphous series according to the composition of the tetrahedral group in the x-ray-structural formulas; he has also constructed graphs, superimposing the imaginary points representing the compositions of the chlorites upon a diagram indicating the parameters $R^{..}O : SiO_2$ and $R^{...}_2O_3 : SiO_2$, showing by lines the axes of the isomorphous chlorite series (Figs. 14 and 15).

The fact that D. P. Serdyuchenko ignored the $Fe^{...} : Fe^{..}$ and $(Fe^{...} + Fe^{..}) : (RO + R_2O_3)$ ratios greatly detracts from the usefulness of this classification.

The classification is based on the following general structural formula for the chlorites, proposed by D. P. Serdyuchenko in 1948 [1953, p. 299]:

$$(R^{..}, R^{...}_{1/3})_{6+\frac{n}{2}} [O_p (OH)_{8-2p}] [Si_{4-n} (Al, Fe^{...})_n] O_{10},$$

where n = 0.....2; p = 0.....2.

The index $(6 + \frac{n}{2})$ in the structural formula corresponds to the composition of the octahedral layers in equivalents of $R^{..}$, $n/2$ being the index of compensation for valence deficiency in the lattice becuase of the substitution of $R^{...}_n$ for part of the silicon (Si_n) in the tetrahedrons.

All varieties of natural chlorites are associated, according to D. P. Serdyuchenko, with the following isomorphous substitutions:

1) $Mg - Fe^{..} - Ni - Mn^{..} - Li$;

2) $Al - Fe^{...} - Cr - Mn^{...}$;

3) $Si - Al - Fe^{...} - Cr$;

4) $Mg_3 - Al_2$; $R^{..}_3 - R^{...}_2$;

5) $O - (2OH)$.

The paper of D. P. Serdyuchenko [1953] contains composite data on x-ray powder photographs of sedimentary and other chlorites.

V. I. Mikheev, who made a special study of the x-ray properties of minerals in the chlorite group, wrote in 1953 that the x-ray data on chlorites to be found in the literature were not accurate enough to serve as standards; these data indicated only three groups of chlorites (magnesian, iron-magnesian, and ferruginous) in a manner similar to that recognized earlier by V. P. Ivanova [1949] on the bases of thermal and optical data.

Brindley and Robinson (symposium on "X-ray methods of identification and the crystalline structure of clay minerals. Chap. 6, Chlorite minerals," 1955) have shown that all chlorites have the same structural framework (Fig. 16), and it is practically impossible to distinguish them on the basis of examination of x-ray photographs alone.

V. I. Mikneev [1957] based his work on the following general formula for the chlorites:

$$Mg_{6-x-y-z} Fe^{..}_z (Al, Fe^{...})_{x+1/3y} \{Al_x Si_{4-x} O_{10}\} [OH]_8.$$

This formula shows the amount of aluminum in the tetrahedral silicon-oxygen layers of the structure that may substitute for silicon. This latter process involves the replacement of bivalent magnesium by trivalent aluminum in the brucite layer. In addition, magnesium may be replaced by aluminum independent of the silicon (y). The formula also takes into account the possibility that bivalent magnesium may be replaced by trivalent iron and that ordinary isovalent isomorphism may hold when magnesium is replaced by iron.

Despite the presence of three types of isomorphism in the chlorites, clear x-ray (powder) photographs have permitted $\underline{x}$ and $\underline{y}$ to be found, and the diagrams prepared by V. I. Mikheev have been used to determine the average dimension of the octahedral cation and of $\underline{z}$; an idea of the chemical composition of the chlorites has thus been obtained.

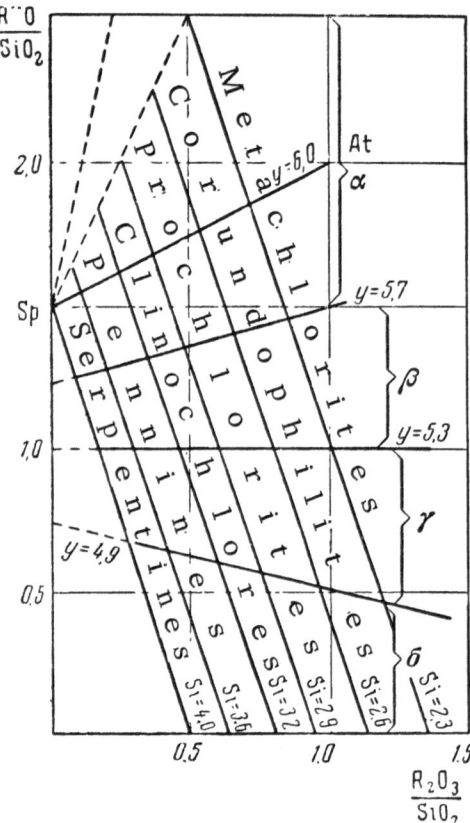

Fig. 14. Classification diagram of chlorites (from Serdyuchenko, 1953) with divisions according to the content of Si in the tetrahedral layers of the lattice and according to the number of octahedral cations, y.

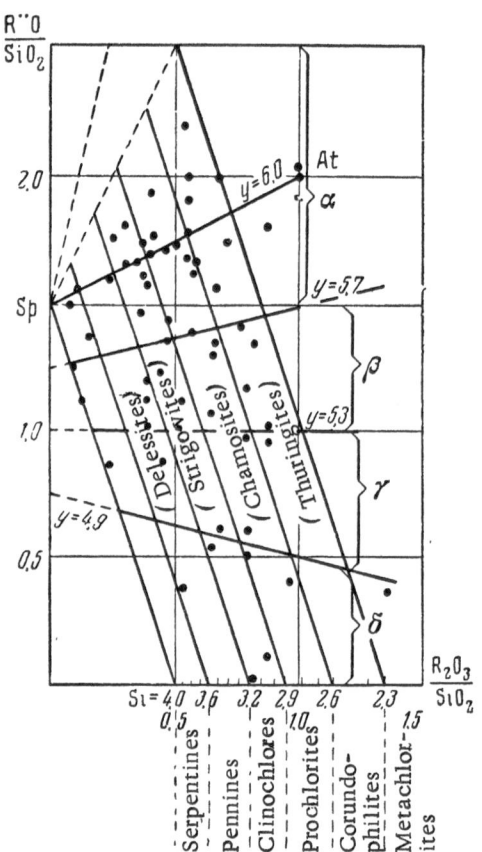

Fig. 15. Classification diagram of chlorites (from Serdyuchenko, 1953) showing the fictive points for chlorites of very different chemical compositions.

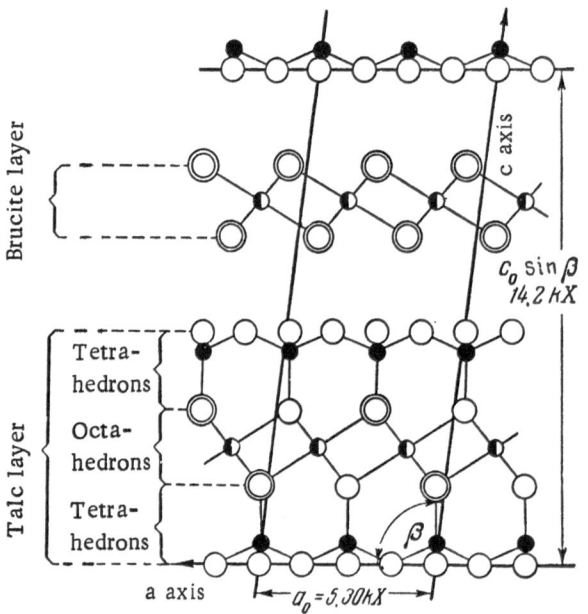

Fig. 16. Structural plan of chlorite on the ac plane (from McMurchy, 1934, and from Brindley and Robinson, 1951).

Iron Chlorites

Iron chlorites are characterized by a considerable content of FeO and Fe_2O_3 or by a high content of iron and a low content of water. Iron chlorites, or ferroferrichlorites (briefly, leptochlorites), like many other iron minerals, indicate the presence or the sequential change of definite oxidation-reduction conditions during diagenesis of sediments or during epigenesis of rocks. Iron chlorites (leptochlorites) are of especial interest during the period of diagenesis of sediments; they may be divided into four principal subgroups: a) ferrous (chamosite, some bavalite, thuringite, and several others); b) predominantly ferrous (most bavalites and thuringites); c) ferrous-ferric (strigovite, cronstedtite, griffithite, and others); d) chiefly ferric leptochlorites, representing distinctive correlatives of glauconite, but containing no potassium (several chlorites from Malka and a number of leptochlorites from the Khalilovo region). In addition, rare discoveries are known of pure ferric chlorites (mackensite). Iron chlorites, or leptochlorites, commonly form entire layers with oolitic structure, occur in the cement of some siltstones and sandstones, and, with other iron minerals (hydrohematite, siderite, and others), form large sedimentary iron-ore deposits (Khalilovo and Malka deposits, Kerch ores, etc.).

The oolites of leptochlorites are relatively homogeneous or consist of concentric layers alternating with ferric oxides (hematite, hydrohematitie, goethite-hydrogoethite), with aluminum oxides (diaspore, hydrargillite, etc.), or with silica; the leptochlorites have clearly precipitated in the form of gel. The concentric structure of the oolites, consisting of leptochlorites with different contents of FeO and Fe_2O_3 or of leptochlorites and hematite-hydrohematite, indicates repeated microfluctuations in the oxidation-reduction boundary (or the pH environment) in the initial sediment [Teodorovich, 1947, 1949, 1951, and others]. For leptochlorites to form it is necessary that the environment in the sediment be such that hydrates of ferric iron, alumina, and silica may react; these conditions apparently must be associated with intense weathering of igneous rocks on the neighboring land [Shvetsov, 1948]. Most of the leptochlorites (three of the four subgroups) have formed under more intensely reducing conditions than glauconite:

a) ferrous leptochlorites were formed in the sulfide-siderite or the siderite mineralogical-geochemical facies, i.e., with the oxidation-reduction boundary somewhat higher than or at the level of (on the average) the surface of the sediment;

b) predominantly ferrous leptochlorites grew in the environment of the siderite facies;

c) ferrous-ferric leptochlorites were also formed during repeated microfluctuations of the oxidation-reduction boundary, but the average position was somewhat below the surface of the sediment. (This mineralogical-geochemical facies has been called by us the ferrous-ferric leptochlorite facies or, tentatively, the leptochlorite facies);

d) only the chiefly ferric leptochlorites formed under oxidation-reduction conditions similar to that of the glauconitic mineralogical-geochemical facies. However, the environments in which leptochlorites and glauconite formed were different, as indicated by the absence of K_2O in the leptochlorites.

However, most glauconite formed in the sea, and glauconite of continental origin has been but recently recognized; whereas leptochlorites are equally abundant in marine and continental subaqueous deposits.

Below are described the principal representatives of the four subgroups of sedimentary iron chlorites, or leptochlorites.

a) Ferrous Leptochlorites

These are represented by chamosite and by some bavalites and thuringites.

Chamosite$-Fe_4^{''} Al[Si_3AlO_{10}][OH]_6 \cdot$ n H_2O or $4FeO \cdot Al_2O_3 \cdot 3SiO_2 \cdot 3H_2O$. The indicated formula is approximate, since the chemical composition of chamosite is variable 34.3-42.3% FeO; 0-6% Fe_2O_3; 22.8-29% SiO_2; 13-20.1% Al_2O_3; 10-13% H_2O; MgO may also be present in considerable quantities. The crystal system is monoclinic. The hardness is 3, the specific gravity 3.0-3.4. The mineral is dark green, greenish dark gray, to black. The refractive index Nm = 1.62-1.66 (generally near 1.64); Ng$-$Np = 0.010-0.012; the mineral is biaxial, negative, and weakly pleochroic. Chamosite is found in oolites and in dense microscopically scaly or cryptocrystalline masses; it is also found in the cement of some sandstones. It is one of the most widespread of the leptochlorites, occurring in many sedimentary iron ores; it forms where there is insufficient oxygen in a littoral or shallow-water environment (also in lagoons and lakes), being confined to deposits of the siderite (chamosite-siderite) and sulfide-siderite (sulfide-chamosite) mineralogical-geochemical facies. Chamosite is also found in the lower parts of present-day peat deposits with beds of lacustrine-paludal calcite. According to D. P. Serdyuchenko [1953] the minerals called chamosite fall into several series in his classification of chlorites.

b) Predominantly Ferrous Leptochlorites

These are represented by most of the bavalites, the thuringites, and the greenalites.

Bavalite. This is an iron chlorite frequently considered to be a variety of prochlorite, $(Mg, Fe)_{4.5}$ · $Al_{1.5}(Al_{1.5}Si_{2.5}O_{10})(OH)_8$, rich in $Fe^{..}$. It is characterized by a distinct predominance of FeO over Fe_2O_3; it is distinguished from glauconite, apart from the structural-textural features (oolites), by the absence of potassium. One chemical analysis gives 21.40% SiO_2, 24.90% Al_2O_3, 8.0% Fe_2O_3, 30.60% FeO, 8.0% CaO; 4.40% MgO, and no H_2O or K_2O. Another analysis of bavalite indicates that, according to the Fe_2O_3:FeO ratio, the mineral belongs to the group of ferrous leptochlorites: 22.27% SiO_2, 0.08% TiO_2, 21.40% Al_2O_3, 0.67% Fe_2O_3, 43.1% FeO, 0.05% MnO, 2.35% MgO, 0.15% CaO, 0.35% $Na_2O + K_2O$, 10.10% H_2O^+, and 0.11% H_2O^-. The mineral is green or dark green. The indices of refraction are Ng = 1.667, Np = 1.658; Ng−Np = 0.009-0.010-0.012. The mineral is pleochroic, the structure lamellar and oolitic.

D. P. Serdyuchenko [1953] has referred bavalite to the clinochlore series with the general formula: $(R^{..}, R^{...}_{2/3})_{6.45}[O_p(OH)_{8-2p}][Si_{3.10}Al_{0.90}]O_{10}$.

Thuringite — $Fe_{3.5}^{..}(Al, Fe^{...})_{1.5}[Si_{2.5}Al_{1.5}O_{10}] \cdot [OH]_6 \cdot nH_2O$, or $7FeO \cdot 3(Al, Fe)_2O_3 \cdot 5SiO_2 \cdot nH_2O$. The indicated formula is approximate, since the composition of the mineral is variable: 19.8-39.3% FeO; 1.3-7.2 to 10.6-31.7% Fe_2O_3; 15.6-25.1% Al_2O_3; 19.4-28.8% SiO_2; 4.6-13.2% H_2O. The system is monoclinic. The hardness is 2.0-2.5, the specific gravity 3.15-3.20. The mineral is green, of various shades, to dark green. The index of refraction is Nm = 1.64-1.68; Ng − Np = 0.005-0.11; the mineral is biaxial, negative, and strongly pleochroic. It is found in crystalline and microscopically flaky masses, rarely in fine-grained flakes.

This mineral has been found in several slightly metamorphosed sedimentary iron-ore deposits; it is commonly associated with magnetite, and locally may be found with later siderite. In some deposits thuringite is hydrothermal. According to D. P. Serdyuchenko [1953], the minerals called thuringite fall into five series in his classification of chlorites.

Greenalite — $Fe_9^{..}Fe_2^{...}[Si_4O_{11}]_2[OH]_{12} \cdot 2H_2O$. A hydrous iron silicate. The indicated formula is approximate. The specific gravity is 2.8. The crystal system is monoclinic. The mineral is black-green, green, or yellowish and brown (oxidation). It is isotropic (ultramicroscopic crystals, giving an x-ray picture), with an index of refraction N = 1.650-1.654, or it is birefringent. It is soluble in hydrochloric acid. It is known from metamorphic (ancient sedimentary) iron-ore formations at Lake Superior (Minnesota, USA).

Three analyses of greenalites are given:

Analysis 1*: 30.08% SiO_2, 34.84% Fe_2O_3, 25.72% FeO, 9.35% H_2O^+;

Analysis 2: 38.00% SiO_2, 8.40% Fe_2O_3, 46.56% FeO, 7.04% H_2O^+;

Analysis 3: 33.58% SiO_2, 11.16% Fe_2O_3, 45.19% FeO, 10.07% H_2O^+.

The greenalite in analysis 3 is birefringent: Ng = 1.685, Np = 1.674; Ng−Np = 0.011.

c) Ferrous Ferric Leptochlorites

This subgroup includes a number of iron chlorites (strigovite, cronstedtite, griffithite, several greenalites, and others) for which the FeO and Fe_2O_3 contents are characteristically of approximately equal value.

---

* The chemical composition of the part dissolved in hydrochloric acid.

Strigovite — $(Fe", Mg)_3 (Fe"', Fe", Al)_2 [Si_3 AlO_{10}][OH]_8$. This mineral is monoclinic (?). Specific gravity is 2.79-3.14. The color is dark green. Refractive indices are Ng = Nm = 1.61-1.67; Np = 1.59-1.65; Ng−Np = 0.010-0.014 to 0.020. For oolitic strigovite from Timan (Table 12), Ng = 1.666, Np = 1.656; Ng−Np = 0.010. The mineral occurs in oolites and in microscopically flaky masses. It decomposes easily in acid (weak hydrochloric acid when heated). It is known to form an essential part of sedimentary hydrohematite-hydrogoethite-leptochlorite ores in the Mesozoic rocks of the Urals (Table 12), and also occurs in cavities in granite.

The Alapaevsk strigovite is considered to be delessite in which almost all the magnesium has been replaced by bivalent iron. Delessite, as is well known, is a term for the ferruginous variety of chlorite (orthochlorite) − pennine − the chemical composition of which is $(Mg, Fe)_5 Al(AlSi_3O_{10}) (OH)_8$.

D. P. Serdyuchenko [1953] has referred the minerals called strigovite to the clinochlore series in his classification of chlorites.

Cronstedtite — $Fe_4" Fe_2"' [Si_2 Fe_2 O_{10}] [OH]_8$, or $4FeO \cdot 2Fe_2O_3 \cdot 2SiO_2 \cdot 4H_2O$. The mineral is monoclinic. The FeO and Fe_2O_3 contents are 32.83-46-86% and 28.74-29.72%, respectively. The hardness is 3.5, the specific gravity 3.34-?.45. The mineral is green-black to brown-black, in thin plates in transmitted light the color is emerald green. The index of refraction is Nm = 1.72-1.80 and the birefringence is high; the mineral is biaxial, negative. It dissolves in hydrochloric acid with the separation of silica gel. It is rarely found with pyrite, pyrrhotite, and siderite. According to D. P. Serdyuchenko the minerals called cronstedtite fall into two series in his classification of chlorites.

Griffithite — $4(Mg, Fe, Ca)O \cdot (Al, Fe)_2O_3 \cdot 5SiO_2 \cdot 7H_2O$. The indicated formula is approximate. According to A. N. Winchell and H. Winchell, griffithite contains 7.32% Fe_2O_3 and 7.83% FeO. The specific gravity is 2.31. The color is dark green. Refractive indices are Ng = 1.572, Nm = 1.565. Np = 1.485; Ng−Np = 0.087; the mineral is biaxial, optically negative, and pleochroic.

d) Chiefly Ferric Leptochlorites

Thus subgroup includes some of the iron chlorites from the Khalilovo Jurassic sedimentary iron-ore deposits and from similar deposits along the Malka River in the Northern Caucasus (Table 13).

These iron chlorites are fine-crystalline, microcrystalline, and cryptocrystalline, and colloidal; they are green, dark green, rarely yellow-green, and, in places, black-green, with metacolloidal structure.

Finally, there are rare silicates of ferric oxide along, for example, mackensite, an iron chlorite with the chemical composition of $Fe_2O_3 \cdot SiO_2 \cdot H_2O$. The color is iron black to greenish black. The mineral is found in chloritic iron ores (in Czechoslovakia).

Iron-Magnesium Chlorites

Metasomatic authigenic chlorites of the iron-magnesium group have now been found in ancient weathering zones. These chlorites, in addition to iron chlorites proper, or leptochlorites, play an important role in the weathering zone, forming during weathering of primary dark minerals (amphiboles, monoclinic pyroxenes, etc.), and of chlorites, themselves being both endogenetic and exogenetic. Hydrochlorites were formed during the latter processes. Magnesian, slightly ferruginous, authigenic chlorites are sometimes observed in the cement of sandstones and in several other sedimentary rocks.

Magnesium and iron-magnesium chlorites (endogenetic and exogenetic) are found in the weathering zone, according to I. I. Ginzburg and I. A. Rukavishnikova [1951]: 1) in various chlorite and talc-chlorite schists; 2) at the contacts of basic dike rocks with serpentinites; 3) in veinlets in pyroxenite dikes; and 4) in decomposed amphibolites, gabbro-amphibolites, and serpentinites. Various nickel-bearing hydrated chlorites occur with the weathering products of nickel-free chlorites (or hydrochlorites).

As noted above, chlorites may be endogenetic or exogenetic. In the zone of weathering chlorite forms in the place of various minerals: diopside, augite, enstatite, uralitic hornblende, tremolite, actinolite, biotite, antigorite, and, in places, zeolites. The chlorite content increases from the lower zone of the weathering profile to the middle zones; it decreases in the upper zones, by change to jefferisite and hydrochlorite at first and then to montmorillonite − halloysite − kaolinite, because of decomposition accompanied by separation of silica.

TABLE 12

Chemical Composition of Strigovites from the Alapaevsk Iron-Ore Deposits [Krotov, 1936; Uspenskii, 1936] and of Strigovite from Southern Timan [Serdyuchenko, 1948]

Sample	SiO_2	TiO_2	Al_2O_3	Cr_2O_3	Fe_2O_3	FeO	MnO	CaO	MgO	K_2O	Na_2O	CO_2	H_2O below 110°	Loss on heating	Totals
I — cement of "subaqueous ore mass" (Alapaevsk region)	34.80	0.72	22.74	0.27	11.99	13.45	0.03	0.73	0.76	0.22		None	3.39	11.10	100.20
II — green "belik"* (Alapaevsk region)	36.11	0.15	23.42	0.07	12.04	12.55	0.02	0.49	0.61	0.44	0.05	0.15	3.48	10.43	100.01
III — greenish gray colitic strigovite (Givetian series in Southern Timan)	28.40	—	12.6	—	18.8	26.2	—	0.1	5.2	—	—	—	—	8.7	100.00

*Tr. note: a local term in the Urals for whitish, fragmental slope wash and aluvial-fan deposits (of Mesozoic age).

TABLE 13

The Chemical Composition of Chiefly Ferric Chlorites

Locality	SiO_2	FiO_2	Al_2O_3	Fe_2O_3	Cr_2O_3	FeO	MnO	NiO	MgO	CaO	Na_2O	K_2O	H_2O+	H_2O-	Totals	Remarks
Khalilov	18.42	0.99	18.92	36.57	0.45	5.40	0.23	0 46	1.03	Trace	—	--	18.40		100.87	—
»	19.55	0.08	0.20	40.45	0.26	15.62	0.07	0.17	3.75	None	—	—	16.91	2.38	99.73	$P_2O_5 = 0.22\%$
»	25.61	0.03	18.56	24.66	0.53	10.08	0.16	0.44	2.38	1.10	0.62		11.71	4.95	100.83	—
»	24.84	0.33	0.00	41.96	0.47	12.15	0.19	2.05	1.03	0.00	—	—	11.57	4.78	99.49	$P_2O_5 = 1.12\%$
»	24.36	—	16.21	23.38	0.43	11.08	0.80	0.66	2.05	0.24	—	—	17.36	—	100.47	—
»	19.21	0.36	17.29	28.53	1.42	14.14	0.39	0 11	1.84	0.15	0.44	0.15	14.96	—	99.16	$CO_2 = 0.17\%$
Malka	29.22	0.29	24.45	13.87	0.30	6.66	Trace	—	6.82	2.44	—	—	8.98	6.64	100.21	$CO_2 = 0.34\%$; Ng — Np = = 0.003 — 0.015
»	29.90	None	21.32	19.14	None	7.90	None	0.85	6.15	0.70	—	—	12.76	0.60	99.68	$CO_2 = 0.08\%$; S = 0.28%

I. I. Ginzburg and I. A. Rukavishnikova [1951] have emphasized the similarity between weathering of feldspars (with gradations to mica, hydromica, and kaolinite) and weathering of pyroxenes, amphiboles, and chlorites (with subsequent formation of jefferisite, hydrochlorite, halloysite, and kaolinite). They point out that exogenetic chlorite begins to form at depth, particularly in the zone of katagenesis, and continues to develop in the zone of weathering (in the lower and middle parts). Pennine, clinochlore, and prochlorite may correspondingly yield hydrochlorite in the zone of weathering; ferruginous varieties may also develop. Chloritization of zeolites near the surface is also known where magnesium and iron are present in aqueous solutions.

Pennine — $(At + Dn)_{20-40} (FeAnt + Dn)_{0-20} (Ant + FeAnt)_{80-60} (Ant + At)_{100-80}$. The mineral is monoclinic. The unit cell contains two mica layers and two brucite layers. Crystals are tabular, pseudohexagonal, occurring in scaly, lamellar aggregates. The hardness is 2.0-2.5, the specific gravity 2.60-2.85. The mineral is green, of various shades, rarely rose-colored, purple, or white. A perfect cleavage parallel to (001) may be observed. The indices of refraction are $Ng = 1.57-1.58$, Nm 1.57-1.58, $Np = 1.57$; $Ng-Np = 0.000-0.004$; the mineral is optically negative or positive, 2V being small (becoming 0° in some varieties); pleochroism is distinct in greens and yellowish greens. The birefringence is characteristically anomalous (dark blue interference color), and flakes are flexible but not elastic. The mineral generally decomposes in sulfuric acid.

Pennine occurs in metamorphic greenstones and in the zone of weathering (as an endogenetic and, partly, an exogenetic mineral), where it may grade into hydrochlorite.

Clinochlore — $(At + Dn)_{40-60} (FeAnt + Dn)_{0-20} (Ant + FeAnt)_{60-40} (Ant + At)_{100-80}$. Clinochlore is monoclinic. Crystals are hexagonal, platy, or tabular, forming coarse-scaly to crypto-scaly aggregates. The hardness is 2.0-2.5, the specific gravity 2.61-2.78. The color is green, of various shades, sometimes yellow. Cleavage is perfect along (001). The indices of refraction are $Ng = 1.57-1.59$, $Nm = 1.56-1.58$ or 1.57-1.58, $Np = 1.56-1.58$; $Ng-Np = 0.004-0.010$; the mineral is positive, generally with a small optic angle, but the angle may range from 0 to 70°. Clinochlore is distinguished from pennine by inclined extinction. It decomposes completely in concentrated sulfuric acid.

Clinochlore occurs in metamorphic greenstones and in the zone of weathering (both as an endogenetic and an exogenetic mineral), where it grades into the corresponding hydrochlorites.

Prochlorite — $(At + Dn)_{60-80} (FeAnt + Dn)_{20-40} (Ant + FeAnt)_{40-20} (Ant + Dn)_{80-60}$. For varieties containing 75% of the amesite and daphnite molecules, the formula is $(Mg,Fe)_{4.5}Al_{1.5}[Al_{1.5}Si_{2.5}O_{10}][OH]_8$. Varieties rich in iron are called bavalite. The crystal system is monoclinic; crystals are pseudohexagonal and form scaly aggregates. The hardness is 1.5-2.0, the specific gravity 2.78-2.86. The mineral is green or blackish green. Cleavage is perfect along (001). The refractive indices are $Ng = 1.60-1.61$, $Nm = 1.59-1.61$, $Np = 1.59-1.60$; $Ng-Np$ is about 0.010, sometimes as low as 0.004. Prochlorite is characteristic of chlorite schists, but more ferruginous varieties are known as hypergene minerals in ancient weathering zone, in the serpentinites of the Malka region for example, where magnesian-ferruginous iron prochlorites have been found [Serdyuchenko, 1953].

"Clay Group" Silicates

This very important group of minerals makes up the argillaceous rocks (clays and mudstones) and is present in marls, argillaceous limestones, and dolomites and in the cement of many sandstones and siltstones. When we recall that argillaceous rocks form approximately 60% of the bulk of sedimentary rocks, the great importance of "minerals of the clay group" becomes very clear. It is true that argillaceous rocks and clay admixtures in other rocks contain both allogenic and authigenic mineral components. For oil geologists argillaceous rocks are of special interest as source rocks for most oil deposits and bituminous accumulations in general and as rocks that have repeatedly acted as impermeable cap rocks beneath which oil has accumulated and has been preserved.

Sedimentary clay minerals, especially authigenic, are commonly found in small particles (too small to be studied under the microscope) or in polymineralic masses; their study requires an entire complex of methods: x-ray analysis (Fig. 17), thermal analysis (thermal curves and dehydration curves, Fig. 18), electron-microscope examination, chemical analysis (complete and partial — for spot tests), ultraviolet microscopy (in part), and spectral analysis. All these techniques, applied to clay material that has been separated and cleansed from impurities and that is preferably monomineralic (for example, particles of a certain size fraction), in addition to immersion studies and thin-section examinations under the microscope, to specific gravity determinations, to tests with dyes, and, occasionally, to determination of adsorptive properties, make it possible to identify precisely clay minerals or mixtures of clay minerals.

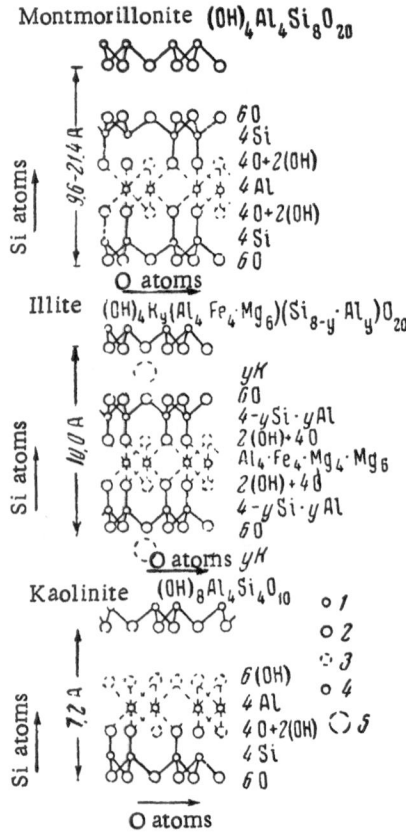

Montmorillonite $(OH)_4 Al_4 Si_8 O_{20}$

Illite $(OH)_4 K_y (Al_4 \cdot Fe_4 \cdot Mg_6)(Si_{8-y} \cdot Al_y)O_{20}$

Kaolinite $(OH)_8 Al_4 Si_4 O_{10}$

Fig. 17. Unit layers of montmorillonite, illite, and kaolinite. 1) Si atoms, 2) O atoms, 3) OH, 4) Al and other atoms, 5) K atoms.

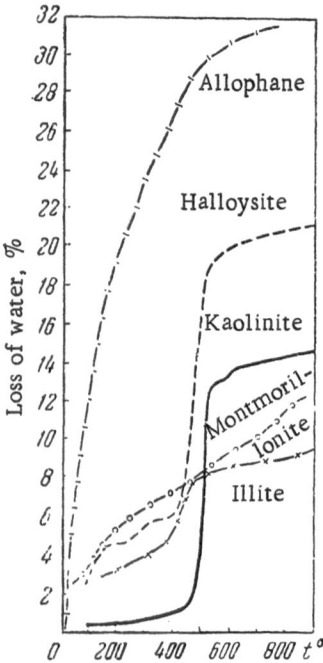

Fig. 18. Dehydration curves of the clay minerals.

Naturally it is impossible and unnecessary, in general, to use all the techniques enumerated above, and, consequently, clay minerals are examined by only two or three methods, in addition to microscopic study, which are sufficient to permit reliable determination of the mineral composition. It should be emphasized that for these analyses, too, material should be carefully selected in the field, the amount of the samples being limited. In determining finely dispersed clay minerals, one should not base his conclusions on a single microscopic examination, since the absorbing properties of many clay minerals make it difficult to determine refractive indices; on the other hand, the optical constants of the clay minerals, which distinguish the subgroups, partially overlap.

The clay minerals, being chiefly hydrous aluminosilicates or silicates of aluminum, may be either microcrystalline or colloidal. They are characterized by scaly or tabular particles and by small size, by generally low indices of refraction (except for iron-bearing varieties), by low hardness, and by low specific gravity (less than quartz). Crystalline minerals of the clay group are monoclinic.

The classification of the minerals in the clay group has not yet been firmly established. We prefer a subdivision of the crystalline clay minerals on the basis of the $Al_2O_3 : SiO_2$ ratio in their formulas [Teodorovich, 1946$_1$]. Five principal groups of clay minerals are distinguished: 1) allophanes or allophanoid minerals (amorphous clay minerals); 2) kaolinites and halloysites; 3) montmorillonites (including beidellites); 4) hydromicas, or illites; and 5) magnesian clays (sepiolites and palygorskites). A sixth group might be added to include monothermites, which apparently represent a series of authigenic minerals intermediate between kaolinite and montmorillonite; but, according to other authors, these are structural intergrowths of kaolinite and hydromuscovite. Sedimentary rocks may also contain hydrochlorites, vermiculites, and several other minerals representing alteration products (frequently weathering products) of chlorites, micas, etc.

This indicated classification agrees in considerable measure with the subdivision of clay minerals on the basis of the $Al_2O_3 : SiO_2$ ratio, particularly:

1) the allophane group normally has an $Al_2O_3 : SiO_2$ ratio of 1 : 1 to 1 : 1.8 (in general the allophane group, consisting of amorphous clay minerals with variable contents of SiO_2, Al_2O_3, and H_2O, may show fluctuations in the $Al_2O_3 : SiO_2$ ratio ranging from 1 : 0.4 to 1 : 2.1);

2) in the kaolinite and halloysite group the $Al_2O_3 : SiO_2$ ratio is 1 : 2;

3) in the monothermite group the ratio is 1 : 2 to 1 : 3;

4) in the hydromica group it is 1 : 2 to 1 : 6;

5) in the montmorillonite group, 1 : 3 to 1 : 6;

a) in beidellite, about 1 : 3, more precisely 1 : 3 to 1 : 3.75;

b) in montmorillonite proper, 1 : 4 to 1 : 6 (in isomorphous mixtures with beta-cerolite):

6) in magnesian clay minerals the ratio is $1:4$ to $0:3$ (for palygorskite, $1:4$; for sepiolite, $0:3$).

Thus, monothermites are naturally described after the kaolinite group but before the hydromica and montmorillonite groups.

The magnesian clay minerals should be considered last, as a group of magnesian aluminosilicates with an $Al_2O_3:SiO_2$ ratio of $1:4$ and as being magnesium silicates with no Al_2O_3.

I. I. Ginzburg and I. A. Rukavishnikova [1951] subdivided the clay minerals in the ancient weathering zones of the Urals, which they studied, in the following way:

I. Halloysites—A. Aluminous; B. Ferruginous (ferrihalloysites).

II. Beidellites—A. Aluminous; B. Ferruginous (ferribeidellites); C. Magnesian.

III. Montmorillonites—A. Aluminous; B. Ferruginous (ferrimontmorillonites); C. Magnesian.

IV. Allophanes—A. Aluminous; B. Ferruginous.

The value of the "clay group" minerals as indicators of reactions in an environment (pH values was considered above (in Chapter 1 of the present paper).

Allophane Minerals, or Allophanoids

Amorphous clay minerals consisting of fine colloidal mixtures of hydrogels of alumina and silica, and, possibly, also containing hydrous silicates of aluminum ($Al_2SiO_5 \cdot H_2O$) in intimate mixtures with colloidal SiO_2 or Al_2O_3; in the indefinite chemical composition, the content of Al_2O_3 is generally greater (in wt. %) than SiO_2, or is approximately the same, and water is abundant.

Minerals in the allophane group are not chemical compounds, but are solid pseudosolutions; it is therefore suitable to examine the entire allophane group (i.e., the allophanoids) as a whole.

They have been described under various terms (allophane, schrotterite, samoite, etc.).

For allophane proper, the formula $Al_2O_3 \cdot SiO_2 \cdot nH_2O$ is given, where $n = 5-6$. When n is 5, the chemical composition of allophane is 23.8% SiO_2, 40.5% Al_2O_3, 35.7% H_2O. The general formula for the allophanoids is $mAl_2O_3 \cdot nSiO_2 \cdot pH_2O$, in which the $Al_2O_3:SiO_2$ ratio normally ranges from $1:1$ to $1:1.8$, but, in general, may range between $1:0.4$ and $1:2.1$.

The Al_2O_3 content ranges from 23.5 to 41.6%, SiO_2 from 18.5-21.4 to 30.9-39.1%, and H_2O from 30.5-39.0 to 41.5-43.9%. Impurities frequently present are Fe_2O_3 (ferruginous allophanes), MgO, CaO, $K_2O + Na_2O$, and P_2O_5; the $RO:SiO_2$ ratio is approximately $1:7$.

F. A. Nikolaevskii [1912] distinguished two principal groups among the allophanoids: a) allophane, when $Al_2O_3:SiO_2 \cong 1:1$; and b) schrötterite, when the ratio of $Al_2O_3:SiO_2 \cong 1:0.5$ or $1:0.4$. In addition, he distinguished varieties of allophanoids with $Al_2O_3:SiO_2$ ranging from $1:1$ to $1:2$, which he proposed to call samoites.

Mixtures of Al_2O_3 and SiO_2 hydrogels are known both with a sharp predominance of aluminum oxide hydrate (shanyavskite,* with $Al_2O_3:SiO_2 = 1:0.05$), grading into alpha-kliachite or aluminogel (p. 136), and with a marked dominance of amorphous silica (opal), in particular "lardite" of P. A. Zemyatchenskii. Solid pseudosolutions of gel of SiO_2 and Al_2O_3 hydrates with a distinct predominance of silica are also known.

The hardness of the allophanoids is normally about 3 (in general, it may range from 2.0 to 3.5), and the minerals are brittle; the specific gravity is normally 1.85-1.89 (in general, it may range from 1.51 to 2.10). The color is white, pale blue, greenish yellow, and, rarely, dirty green or brown. Allophanoids are found in glass-like masses with semiconchoidal or conchoidal fracture, as crusts with nodular surfaces, and rarely, in earthy or powdery masses. The minerals are isotropic. The index of refraction ranges from 1.468 to 1.512, most commonly being about 1.480. Friable earthy varieties of allophane commonly contain free alumina hydrate.

Allophanes are distinguished by their glassy appearance, by their transparency or semitransparency, and by their great friability. They give a thermal curve typical of amorphous substances in which water is present in solution. The thermal curves show a gradual elimination of this water, an endothermic effect below 200°, but at

---

*With 53.55% Al_2O_3, 1.33% SiO_2, and 40.95% H_2O, i.e., a composition corresponding to the formula $Al_2O_3 \cdot 0.05SiO_2 \cdot 5.48H_2O$.

900-1000° there is an exothermic effect corresponding to the crystallization of the amorphous material. Allophanes decompose easily in 10-15% hydrochloric acid; in the process the alumina is completely extracted and gelatinized silica remains. Precise identification of the allophanoids requires the use of chemical analysis.

The allophanes are exclusively exogenetic minerals; they are commonly present in the zone of weathering, in the oxidized zones of ore deposits, and in fractures or cavities in sedimentary rocks; not infrequently they are found with halloysite, occasionally with chrysocolla.

In carbon dioxide waters the minerals suffer carbonatization.

Kaolinite and Halloysite Group — $Al_2O_3 \cdot 2SiO_2 \cdot (2-4)H_2O$

Kaolinite Subgroup

This group contains three minerals: kaolinite proper (the most widespread), dickite, and nacrite.

Kaolinite — $Al_4[Si_4 \cdot O_{10}][OH]_8$, or $Al_2O_3 \cdot 2SiO_2 \cdot 2H_2O$. A hydrous aluminosilicate, or basic silicate of aluminum, crystallizing in the monoclinic (or triclinic) system. The chemical composition is 39.5% Al_2O_3, 46.5% SiO_2, and 14% H_2O. The hardness is 1.0-2.5, the specific gravity 2.58-2.63. The mineral is white, grayish white, yellow, occasionally brownish. Cleavage is perfect along (001). The indices of refraction are Ng = 1.565-1.567, Nm = 1.562-1.565, Np = 1.559-1.561; Ng−Np = 0.006-0.007; the mineral is biaxial and negative.

Kaolinite has a sheet structure, each layer consisting of one silicon-oxygen tetrahedral unit and one aluminum-oxygen octahedral unit, combined into a single sheet so that the tips of the octahedrons adjoin the tips of the tetrahedrons (Fig. 19). Kaolinite is the principal constituent of kaolins.

Dickite. This mineral is monoclinic. The indices of refraction are Ng = 1.566, Nm = 1.562, and Np = 1.560; Ng−Np = 0.006. In contrast to kaolinite and nacrite, dickite is optically positive. A comparison of the structures of the crystal lattices of kaolinite and dickite is given in Fig. 20.

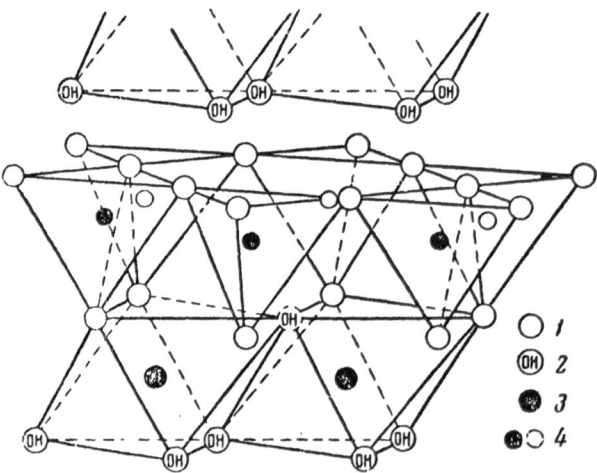

Fig. 19. Diagrammatic sketch of the structure of the kaolinite layer. (After Grüner, 1932.) 1) O atoms, 2) OH, 3) Al atoms, 4) Si atoms.

Nacrite. The crystal system of nacrite is monoclinic. The indices of refraction are Ng = 1.563; Nm = 1.562, Np = 1.557; Ng−Np = 0.006; the sign is negative, occasionally positive.

Kaolinite, dickite, and nacrite are differentiated by their extinction angles: in kaolinite this angle is 1-4° (Np∧⊥(001), but Nm∧a = 0°; in nacrite Nm∧a = 10-12°; in dickite Np∧c ~ 20°, and Nm∧a = 11°.

The minerals of the kaolinite subgroup are characteristically scaly, the refractive indices are higher than that for Canada balsam (positive relief, especially when the iris diaphragm is partially closed), and the structure is sometimes vermicular, easily recognized in thin section; the determination of the optical constants aids in identifying the minerals. The most reliable identifications, especially in cryptocrystalline masses, is attained by means of thermal curves, x-ray measurements, and electron micrographs. The principal groups of clay minerals, including the kaolinite subgroup, may also be distinguished by use of dyes: spectrophotometric analysis [Vedeneeva and Vikulova, 1952, 1956].

The thermal curve of kaolinite shows two main effects: an endothermic effect with a maximum near 550-600° and an exothermic effect at 950-1000°. The endothermic effect is due to loss of hydroxyl ions and the later development of amorphism in the initial material. The exothermic effect is apparently due to the formation of mullite, $Al_4[Al_4(Si_3Al)O_{20}]$. Finally, a small exothermic peak is noted at 1200°, corresponding to the formation of cristobalite. The thermal curve of dickite differs from this in that the endothermic effect occurs at a somewhat higher temperature, the maximum being found in the interval 600-700°. Nacrite shows two endothermic effects.

Minerals of the kaolinite subgroup are almost unaffected by hydrochloric or nitric acids, but in warm sulfuric acid they decompose rather easily; however, long boiling is necessary for complete decomposition. The minerals of this subgroup are not decomposed in weak hydrochloric acid.

Anauxite. A mineral of the kaolinite subgroup, rich in alumina, with a low water content and with $Al_2O_3 : SiO_2 = 1 : 2$ to $1 : 2.8$.

According to Kerr and Gruner, anauxite is kaolinite in which the aluminum is partially replaced by silicon, a reaction accompanied by a corresponding replacement of the hydroxyl group by oxygen.

Pholerite. A mineral variety relatively rich in aluminum, with $Al_2O_3 : SiO_2$ from $1 : 1.7$ or $1 : 1.8$ to $1 : 2$. It is said that part of the silicon of kaolinite is replaced by aluminum.

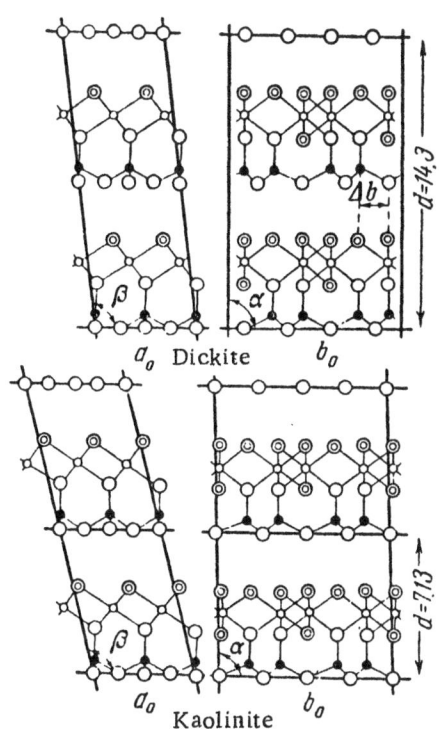

Fig. 20. A comparison of the structures of kaolinite and dickite. (From the paper of Brindley, 1951. See also "X-ray Methods . . .," 1955). Two units of kaolinite and one of dickite are shown. The heavy lines delineate the unit cells. The kaolinite cell is disposed in a single layer.

* * *

Kaolinite is generally of exogenetic surficial origin, forming during the weathering of aluminosilicates in igneous, metamorphic, and, occasionally, sedimentary rocks; nacrite and dickite are chiefly low-temperature hydrothermal minerals; however, hydrothermal kaolinite is known, and nacrite and dickite are found as surface weathering products.

Kaolinite is commonly a characteristic component of the light fraction in sedimentary rocks. It is derived normally from the surface weathering of acid and intermediate igneous rocks and gneisses, but it is occasionally authigenic, forming in an acid environment.

It forms the whole or the greater part of the primary kaolins in ancient weathering zones and of reworked kaolins or kaolinitic clays. Near ancient erosion surfaces montmorillonitic clays may be observed changing to kaolinite [Belyankin and Petrov, 1950]. On the other hand, kaolinite is also formed by precipitation from solutions circulating through sedimentary rocks or saturating sedimentary rocks; this process is commonly accompanied by the replacement of other minerals.

Halloysite Subgroup

In addition to halloysite proper, the halloysite subgroup includes matahalloysite and ferrihalloysite.

Halloysite — $Al_4[Si_4O_8][OH]_{12} \cdot 2H_2O$, or $Al_2O_3 \cdot 2SiO_2 \cdot 4H_2O$. The chemical composition is 34.7% Al_2O_3, 40.8% SiO_2, 24.5% H_2O. Half the H_2O occurs in the mineral as hydroxyl, but the other half is present as the water molecule. The crystal system is monoclinic. The hardness is 1-2; the specific ranges from 2.0 to 2.2-2.4. The average refractive index of halloysite ranges from 1.532 to 1.550 (most commonly 1.543-1.544), increasing with a decrease in water content. The mineral is ordinarily almost isotropic, but occasionally it shows very weak birefringence; the birefringence increases with increase of isomorphous Fe_2O_3.

The structure of halloysite, as proposed by Hendricks and Jefferson, is shown in Fig. 21; the differences between this structure and that of kaolinite are shown in Fig. 22 (after Bates et al.).

A number of characteristic features of halloysite are the ordinarily very low birefringence (almost isotropic, by which it may be distinguished from kaolinite); the lower (than kaolinite) average index of refraction; and the lower specific gravity. The mineral is partly decomposed in acids and alkalies, especially during heating. It is most accurately identified by x-ray measurements, differential thermal analysis, and by electron micrography. A characteristic tubular aspect of halloysite crystals has been recognized under the electron microscope.

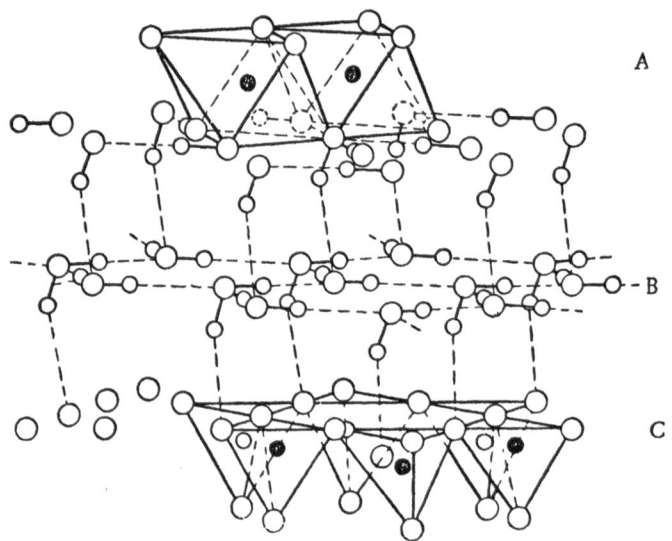

Fig. 21. Diagrammatic sketch of a part of the halloysite $4H_2O$ structure showing a single layer of water molecules. (After Hendricks and Jefferson, 1938.) A and C represent silica tetrahedral layers in halloysite; B is a layer of water molecules.

On heating, halloysite is dehydrated and changes to metahalloysite at 60-110°.

Metahalloysite — $Al_4[Si_4O_{10}][OH]_8$ or $Al_2O_3 \cdot 2SiO_2 \cdot 2H_2O$. The average index of refraction of metahalloysite is 1.549-1.555. The mineral is found in associated with halloysite. The thermal curve of metahalloysite is similar to the thermal curve of kaolinite (a distinct endothermic effect with a maximum near 550° and an exothermic effect at 950-990°; in addition a very small endothermic effect is observed at 105-150°).

Halloysite and metahalloysite are present in a number of clays, in marine muds, in weathering zones, and in kaolin deposits of various types. I. I. Ginzburg and I. A. Rukavishnikova [1951] have noted that in zones of weathering halloysite generally forms during decomposition of plagioclases; it also forms by decay of hornblende and chlorite.

Ferrihalloysite, or iron halloysite — $(Al, Fe)_4[Si_4O_8][OH]_{12} \cdot nH_2O$, or $(Al, Fe)_2O_3 \cdot 2SiO_2 \cdot (n + 2)H_2O$. The specific gravity of ferrihalloysites ranges from 2.04 to 3.68. The minerals are almost isotropic or show weak aggregate polarization. The refractive indices are variable: in the lightest varieties (green, yellow-green) the index is low (1.530-1.550), but in the darker iron varieties it increases to 1.575 and even to 1.614-1.669. The Fe_2O_3 content ranges from 6 to 18% and more. Birefringence is distinct in varieties containing a considerable isomorphous admixture of Fe_2O_3 or that grade into nontronite and montmorillonite. The birefringence ranges from 0.007 to 0.020-0.027. The thermal curves of ferrihalloysite are characterized by four thermal effects: 1) an endothermic effect at 90-155°, corresponding to the elimination of low-temperature water; 2) an endothermic effect with a maximum near 540-590°, representing the loss of hydroxyl ions; 3) an exothermic effect at a temperature near 870-960°; and 4) an exothermic effect near 1080-1160°.

Ferrihalloysites are found in ferrihalloysitic clays, and also (rarely) in two pure varieties: dense with planar-conchoidal fracture and earthy.

Monothermite Group

<u>Monothermite.</u> This mineral of the clay group is chemically similar to kaolinite but is distinguished from that mineral by its content of 2-3 to 5% alkali (especially K_2O) and alkaline earths (chiefly CaO), by a much greater content of water (which is driven off at 150-200° and at 500-550°, and by a greater content of silica ($Al_2O_3 : SiO_2 = 1 : 2$ -1 : 3, more precisely 1 : 2.1 to 1 : 2.92). The hardness is low, the specific gravity 2.6. The average index of refraction is N = 1.55-1.57; Ng = 1.562-1.572-1.576; Np = 1.541-1.551-1.553 (lower than in kaolinite); Ng−Np = 0.013-0.020-0.030, the value increasing with increase in alkali content. Extinction is parallel or almost parallel. The mineral occurs in matted microscopically fibrous (micro-scaly) masses and aggregates.

Monothermite is distinguished from kaolinite by its higher birefringence, a higher content of K_2O, and by its characteristic thermal curve (with only a single endothermic effect, having a maximum at 550°; the exothermic reaction at 900-950° is absent, or almost absent); the mineral was given its name, monothermite, because of this property [Belyankin, 1938]. It strongly absorbs alkalies and dyes. It dissolves in acids more easily than kaolinite. It is distinguished from hydromicas by lower refractive indices and by a greater content of water.

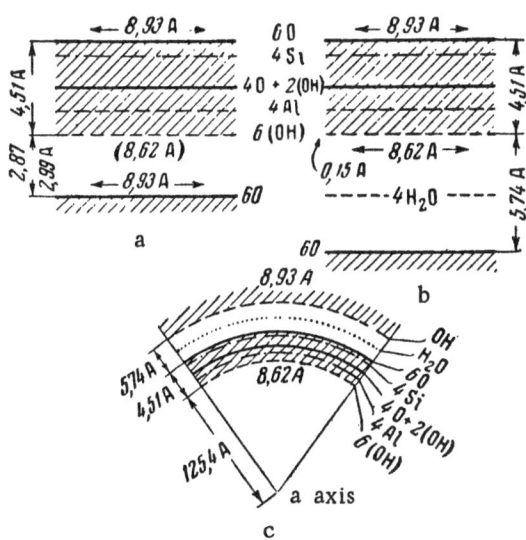

Fig. 22. Diagrammatic representation of the structures of kaolinite and halloysite $4H_2O$. (After Bates, et al., 1950.) a) Distribution of the layers kaolinite; b) distribution of the layers in halloysite (after Hendricks); c) distribution of the layers in halloysite as proposed by Bates et al.

Monothermite consitutes the bulk of a number of highly plastic refractory clays.

According to V. P. Petrov, monothermite is a mineral that forms in a basin during diagenesis of sediments; i.e., it is a typical authigenic syngenetic mineral in sedimentary rocks.

As accumulated data have shown, monothermites represent a mineral series intermediate between kaolinite and montmorillonite (as suggested by M. F. Vikulova) or between kaolinite and hydromuscovite, apparently having a thin-layered structure of the space lattice; they represent a type of "mixed crystals" or approximately mixed-layer structural intergrowths.

The degree of development of the monothermic character, enrichment in K_2O and CaO, may serve as a correlation criterion.

According to V. P. Petrov [1956], the invariable presence of CaO in monothermites (1.5-2.0%) forbids referring the minerals to the hydromica group, since this group, like micas in general, is characteristically calcium free.

<u>Leverrierite.</u> A clay mineral, a hydrous aluminosilicate, commonly referred to the hydromicas, specifically to the hydromuscovite subgroup. In composition leverrierite is similar to kaolinite, but it has a higher birefringence and it contains alkalies and alkaline earths. Termier, who named the mineral, considered it to be muscovite rich in water (approximately 10-15%) and very low in potassium.

The chemical composition of leverrierite is 46.4-49.9% SiO_2, 34.4-38.4% Al_2O_3, 0-3.65% Fe_2O_3, 0-0.44% MgO, a trace to 4.53% CaO, approximately 1.13% K_2O, and 8.65-15.67% H_2O.

The molecular ratios $Al_2O_3 : SiO_2 : H_2O$ range from 1 : 2.0 : 2.2 to 1 : 2.3 : 1.3 or 1 : 2.3 : 2.2.

Termier considered the best analysis and obtained a formula with the ratios $Al_2O_3 : SiO_2 : H_2O = 1 : 2.3 : 1.3$, with a K_2O content of 1.13%.

The hardness of leverrierite is 1.5, the specific gravity 2.5. The mineral characteristically forms a vermicular aggregate (distorted cylindrical prisms) of normal scaly structure, consisting of thin lamellae with hexagonal outline. Leverrierite is distinguished from kaolinite by a higher birefringence, ranging from 0.008-0.011 to 0.020-0.028; for the latter values Ng = 1.582 and Np = 1.554, but in the forms with lower birefringence the indices are lower. Leverriertie is generally found in sedimentary rocks in parallel-tabular vermicular intergrowths, with Ng−Np = 0.008-0.011; it has been described from the Cretaceous and Carboniferous deposits of France.

As may be seen, leverrierite is similar to kaolinite, to hydromuscovite (white hydromica), and to monothermite. According to new data, leverrierite is a mineral of the "mixed crystal" type, consisting of alternating (in a space lattice) layers of oriented structural intergrowths of kaolinite and muscovite.

Hydromica Group

This group includes mica-clay minerals, representing intermediate alteration products of a series of minerals from mica to kaolinite, derived from the hydrolytic decomposition of micas. Purely sedimentary subaqueous, microscopically scaly authigenic minerals of similar composition (monothermite, leverrierite, etc.), are included in this group. Among typical hydromicas one may distinguish white (the most widespread), brown, and green varieties. Some authors designated the hydromicas by numbers, No. 1 being the initial mica and No. 100 being kaolinite. However, this numbering system is generally used only for the series muscovite−hydromuscovite− kaolinite.

White Hydromica. This term represents the mineral varieties that occupy the intermediate position (according to chemical, optical, and other properties) between white mica and kaolinite. The minerals are the products of hydrolysis of white mica, generally muscovite (sometimes paragonite), in which R_2O has been replaced by molecules of H_2O in great part, or almost completely; K is generally replaced by hydronium (H_3O). We have therefore given a diagrammatic sketch of the structure of muscovite (Fig. 23).

The formula of muscovite is $KAl_2(AlSi_3O_{10})(OH)_2$ or $K_2O \cdot 3Al_2O_3 \cdot 6SiO_2 \cdot 2H_2O$; the chemical composition is 11.8% K_2O, 38.5% Al_2O_3, 45.2% SiO_2, and 4.5% H_2O. The crystal system is monoclinic. The specific gravity is 2.76-3.10. The indices of refraction are Ng = 1.588-1.615, Nm = 1.582-1.611, Np = 1.552-1.572; Ng−Np = 0.036-0.040.

Finely ground muscovite produces two endothermic effects with maximums at 850-900° and at about 1100°, the first corresponding to the elimination of constitutional water, the second to the destruction of the crystal lattice of muscovite and to the formation of mullite [Tsvetkov and Val'yashikhina, 1956].

Hydromuscovite − $K_{<1} Al_2[(Si, Al)_4O_{10}] \cdot [OH]_2 \cdot nH_2O$. The indicated formula is approximate.

The term represents a series of minerals in the hydromica group, distinguished from muscovite by a lower content of K_2O (down to 6 or even 3-2%) and a greater quantity of H_2O (up to 8-9%). Hydromuscovite, like other hydromicas, crystallizes in the monoclinic system. The specific gravity and indices of refraction of hydromuscovite are lower than those for muscovite (normally N = 1.55-1.58); these values decrease with increase in content of H_2O.

The thermal curves for hydromuscovite (Tsvetkov and Val'yashnikhina, [1956] are characterized by 1) a double endothermic effect below 250°, 2) an intense endothermic effect at 500-700°, and 3) an endothermic effect, sometimes grading into an exothermic effect (at 1050-1150°). The first two reactions correspond to the loss of low-temperature and constitutional water, the last to destruction of the crystal lattice.

Sericite is referred, in part, to hydromuscovite, but it is, in part, merely fine, flaky muscovite. Hydromuscovite, generally or mostly, is the product of partial hydrolysis of muscovite. It is found in many eluvial and aqueous sedimentary clay rocks. It may apparently be considered an intermediate product during the conversion of feldspars to kaolinite. Hydromuscovite is known in soils that develop on acid and intermediate igneous rocks.

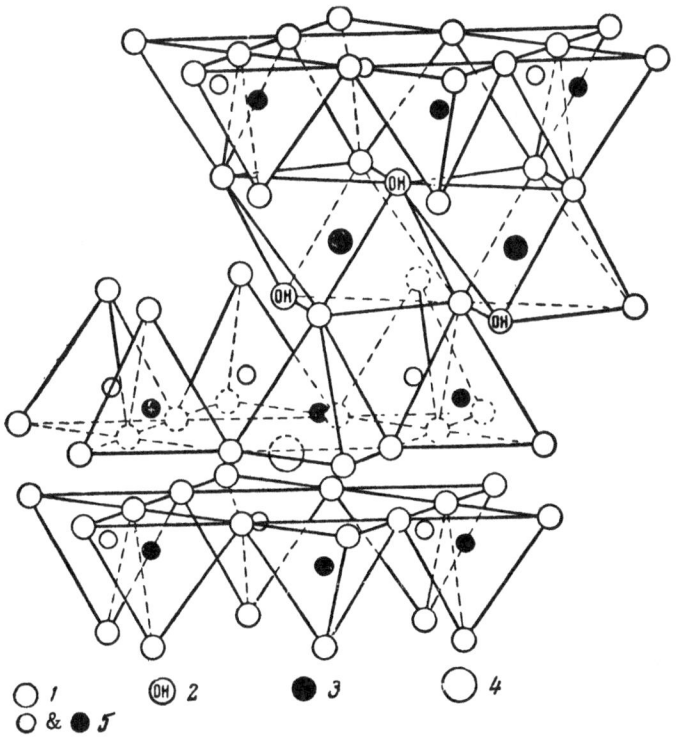

○ _1_   ⊙ _2_   ● _3_   ◯ _4_
◯ & ● _5_

Fig. 23. Diagrammatic sketch of the structure of muscovite. 1)
Oxygen, 2) hydroxyl, 3) aluminum, 4) potassium, 5) silicon (one-
fourth replaced by aluminum).

Illite. This is a clay mineral of the hydromuscovite and hydrobiotite group, described from Paleozoic clays of the USA (state of Illinois), and later from other countries. The optical properties of illite are generally sim- ilar to those for muscovite: Ng = 1.588-1.598; Ng−Np = 0.033. X-ray photographs of illite are commonly sim- ilar to those for muscovite, but they differ in lacking individual lines (or many lines) of muscovite, and the lines are less distinct. In some examples the x-ray photographs are very different.

The thermal curves of illite are clearly different from those for muscovite; three endothermic effects ap- pear on the graphs, with maximums at approximately 130-160°, 550-600°, and 860-900°.

Initially "illite" was a term proposed for all micaceous clay minerals (the alteration products of micas, both muscovite and biotite). Most geologists use it in this way at the present time; others use it chiefly in reference to hydromuscovite clay minerals, distinguishing them from muscovite by a lower potassium content, a greater water content, and lower indices of refraction (1.59-1.55). It is improper to consider illites as only hydromuscovite, since the first descriptions of illites were of altered biotite.

The principal element in the structure of illites is a layer consisting of two silicon-oxygen tetrahedral units and one central octahedral sheet (see Fig. 23). The tips of the tetrahedrons in each silica sheet point toward the center of the structural unit and are combined with the octahedral sheet in such a way that a single layer is ob- tained by suitable replacement of OH by O. The structure of mica and hydromica is similar to the structure of montmorillonite, but it is distinguished by the replacement of some silicon atoms (normally one-fourth the num- ber of silicon atoms) by aluminum; this produces a negative charge that is balanced by potassium ions.

Hydrobiotite. This is a mineral of the hydromica group, distinguished from biotite by a lower content of K_2O, MgO, FeO, Fe_2O_3, Al_2O_3, and SiO_2; an essential constituent part of it is molecular water.

The formula of biotite is $K(Mg, Fe)_3(Si_3AlO_{10})(OH, F)_2$ or $K_2O \cdot 6(Mg, Fe)O \cdot Al_2O_3 \cdot 6SiO_2 \cdot 2H_2O$. The chemical composition varies within the following limits: 6.18-11.43% K_2O, 0.28-28.34% MgO, 2.74-27.60% FeO, 0.13-20.65% Fe_2O_3, 9.43-31.69% Al_2O_3, 32.83-44.94% SiO_2, 0.89-4.64% H_2O, and 0-4.23% F. The crystal system is monoclinic. The specific gravity is 3.02-3.12. The refractive indices of biotite are Ng = 1.60-1.66, Np=1.56- 1.60; Ng−Np = 0.040-0.060.

According to A. I. Tsvetkov and E. P. Val'yashikhina [1956], the thermal curves for biotite (that has been finely cut-ground) are clearly marked by two reactions: 1) an exothermic effect in the form of a gentle undulation between 550 and 900°; and 2) an endothermic effect between 1100 and 1200°. The first effect is due to the oxidation of iron (FeO) and the second to destruction of the mica crystal lattice. The approximate formula for hydrobiotite is $K_{<1}(Mg, Fe)_3[(Si, Al)_4O_{10}][OH]_2 \cdot nH_2O$. X-ray data indicate that hydrobiotite has alternating layers of mica structure and vermiculite structure. The formula for vermiculite is $(Mg, Fe\cdot\cdot, Fe\cdot\cdot\cdot)_3(Si, Al)_4O_{10}]$ $[OH]_2 \cdot 4H_2O$. Hydrobiotie is generally the same color as biotite, but the specific gravity is less (2.78-2.79) and the indices of refraction are lower (Ng = Nm, approximately 1.582, and Np $\cong$ 1.545).

A. I. Tsvetkov and E. P. Val'yashikhina [1956] consider the following thermal effects (or groups of effects) to be characteristic for hydrobiotites and vericulite: 1) a double or triple endothermic effect of low-temperature dehydration below 400° (with successively diminishing maximums at temperatures of about 150, 220, and 360°); 2) an endothermic effect at 800-1000°, appearing different for different samples; and 3) an endothermic effect between 1000 and 1100°, almost absent on some analyses, on others grading into an exothermic effect, and, on one graph, accompanied by another endothermic effect between 1100 and 1200°. The low-temperature group of effects corresponds to the elimination of low-temperature water: below 200° hygroscopic water is driven off; between 200 and 400-500° interlayer water is expelled. At 800-1000° high-temperature combined water (hydroxyl, or constitutional) is eliminated. The succeeding endothermic effect at 1000-1100° apparently corresponds to destruction of the mica crystal lattice.

V. P. Petrov [1948] has stated that the biotite—hydrobiotite—vermiculite series is analogous to the muscovite — hydromuscovite—kaolinite series.

Hydrobiotite has been found in weathering zones on biotite-bearing rocks at places where the alteration products may be carried into aqueous sedimentary deposits and may be subjected to further hydrolysis; the mineral may also be hydrothermal.

Vermiculite — $(Mg, Fe\cdot\cdot, Fe\cdot\cdot\cdot)_3 [(Si, Al)_4O_{10}] [OH]_2 \cdot 4H_2O$. A mica-like mineral with a chemical composition of 37-42% SiO_2, 10-13% Al_2O_3, 5-17% Fe_2O_3, 1-3% FeO, 14-23% MgO, and 8-18% H_2O; in addition, up to 5% K_2O may be present. The crystal system is probably monoclinic. The hardness is 1.0-1.5, the specific gravity 2.4-2.7. The color is brown, yellowish brown, golden yellow, bronze-yellow, and green; shades of green are sometimes observed. The indices of refraction are Ng = Nm = 1.545-1.586, Np = 1.525-1.561; Ng—Np = 0.02-0.03. Biotite grades through hydrobiotite to vermiculite, the hydrobiotite containing alterations of biotite and vermiculite layers in the lattice (according to x-ray data).

A distinguishing feature of vermiculite (to differentiate it from biotite and chlorite) is its strong tendency to swell and to split during heating, before the blow pipe, for example, and to form vermicular, distorted columns and filaments.

Vermiculite is a mineral of the weathering zone, or is an alteration of biotite under other conditions; it is more frequently found in hydrothermally altered (at low temperatures) biotite (or phlogopite) veins and bodies that have developed metasomatically from ultrabasic rocks.

* * *

Hydromicas are distinguished from the initial micas by lower specific gravity, by smaller birefringence, by lesser indices of refraction (chiefly Ng and Nm), and by the nature of the thermal curves. For example, the thermal curves for hydromicas in the hydromuscovite series manifest the following effects: 1) an endothermic effect with a maximum at 120-160°, due to the loss of adsorbed water; 2) a second, normally the chief, endothermic effect, with a maximum at 550-650°, resulting from the partial destruction of the lattice because of expulsion of the hydroxyl group (this loss begins at 400-450°); 3) a third, small endothermic effect at 850-900°, reflecting destruction of the lattice; 4) a final, small exothermic effect at 900-1000°, associated with the formation of spinels from the decomposition products of mullite. The difference between the thermal curves of hydromicas of the hydromuscovite series and the curves of montmorillonites, including beidellite, is found in a lower value for the first endothermic effect and in a greater value for the second endothermic effect. In comparison with the thermal curves for the initial micas, hydromicas present a similar picture. Hydromicas are generally easy to distinguish from other clay minerals under the electron microscope.

Let us examine some other views concerning hydromicas, for example that of I. I. Ginzburg and I. A. Rukavishnikova, who at first also noted that "hydromicas" are normally the products of a stage-like alteration of "micas" (muscovite, sericite, biotite, phlogopite, and, sometimes, chlorite) to kaolinite [1951, p. 487]. Farther on, in the same paper, however, they propose that hydrobiotites, vermiculites, jefferisites, hydrochlorites, hydromuscovites, and hydromicas be differentiated, but within the following frameworks [1951, p. 488].

Hydrobiotites are biotites and phlogopites in which part of the alkali has been removed and has been replaced by hydronium (H_3O), whereas magnesium has been but slightly leached away.

Vermiculites are also biotites and phlogopites, but from which the alkali has been entirely, or almost entirely, removed and magnesium has been but slightly leached away.

Jefferisites are chlorites in the incipient stage of decomposition, from which only a small part of the MgO has been leached away.

Hydrochlorites represent the succeeding stages of decomposition of chlorites, being distinguished by a marked decrease in the MgO content and by an increase in the Al_2O_3 content.

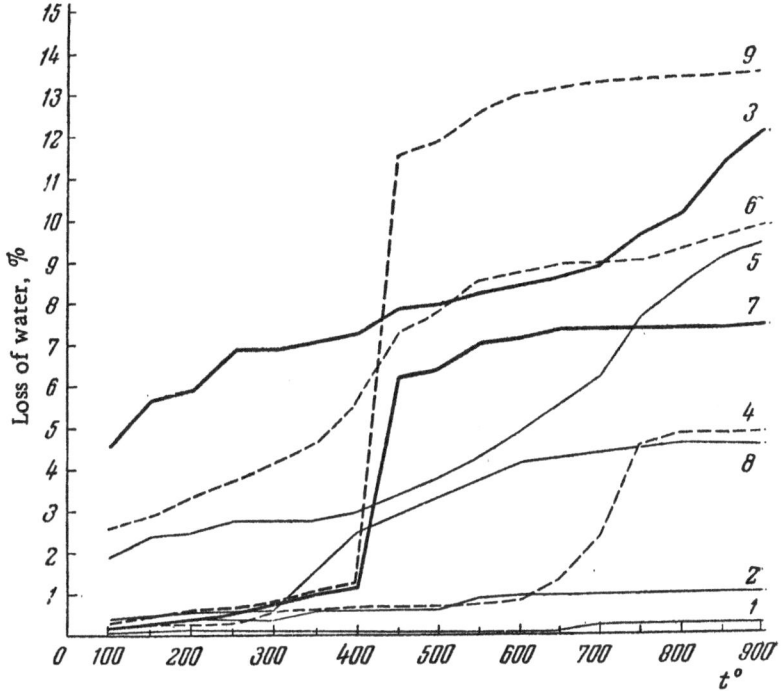

Fig. 24. Dehydration curves of micas, hydromicas, and kaolinite. 1) Phlogopite, 2) biotite, 3) hydrobiotite, 4) muscovite, 5) sericite, 6) illite, 7) hydromica, 8) hydromicaceous clay, 9) kaolinite. (From the paper of I. I. Ginzburg and I. A. Rukavishnikova, 1951.)

Hydromuscovites represent the first stages of weathering of muscovite, the optical constants and the thermal effects still being characteristic of micas (replacement of alkali by hydronium—H_3O—up to 20-25%; the number of hydroxyls in the hydromuscovites being approximately two, which is normal for micas).

Hydromicas represent the succeeding stages of weathering of muscovite and other micas, a rather considerable part of the alkali having been removed, the birefringence having been sharply lowered, and the thermal curve manifesting a well-defined kaolinitic endothermic effect (alkali has been replaced by hydronium up to 25-60%; the number of hydroxyls in the hydromicas is greater than two).

Since the final decomposition product of biotite, muscovite, and chlorite may commonly be kaolinite, all these may grade into hydromicas in the course of their alteration (Fig. 24).

The above discussion clearly shows that the terminology of the mica-like clay minerals has not yet been firmly established ("hydromuscovite," "hydrobiotite," and similar terms are variously understood), but the nature of the genetic processes in the step-like decomposition of the micas and other minerals in the course of altering to kaolinite has been adequately explained. V. P. Petrov, for example, identifies hydromicas with illites.

Hydromicaceous clay minerals are widespread in nature: authigenic minerals have long been known from the clay products of surface weathering of rocks; in subaqueous deposits hydromicas may be allogenic or authigenic (replacing other clay material).

Montmorillonite Group

This group of monoclinic or pseudo-orthorhombic minerals includes varieties of montmorillonite (magnesian, calcium-magnesian, calcium, etc.) and beidellite (i.e., argillaceous varieties) and ferruginous varieties (ferrimontmorillonite and ferribeidellite); recently highly magnesian varieties have been distinguished (magnesian montmorillonite, magnesian beidellite) in addition to other varieties (Table 14). The subgroup of argillaceous varieties (i.e., montmorillonite and beidellite proper) frequently goes under the term bentonite. Ferruginous montmorillonites and beidellites (i.e., ferrimontmorillonites and ferribeidellites) were recognized by I. I. Ginzburg [1936]; they are called "nontronites."

Purely argillaceous montmorillonites include the well-known Fuller's earth (gumbrine), askanite, and other clay varieties. Montmorillonites proper commonly form from volcanic rocks, tuffs, and ashes in hydrous-aluminum-silicate weathering zones (of the montmorillonite type) and on the floors of various basins. Favorable environments for the formation of montmorillonites range from markedly alkaline, through moderately alkaline to slightly alkaline and neutral (generally strongly alkaline for Ca-montmorillonite, moderately alkaline for Ca-Mg-montmorillinite, and slightly alkaline or neutral for Mg-montmorillonite).

TABLE 14

Structural Formulas for Minerals of the Montmorillonite Group (After I. I. Ginzburg and I. A. Rukavishnikova, 1951)

Mineral	Formula
Montmorillonite	$Al_2[Si_4O_{10}][OH]_2 \cdot nH_2O$
Ferrimontmorillonite	$(Al, Fe)_2[Si_4O_{10}][OH]_2 \cdot nH_2O$
Beidellite	$Al_2[Al_1Si_3O_{10}][OH]_2 \cdot nH_2O$
Ferribeidellite	$(Al, Fe)_2[Al_1Si_3O_{10}][OH]_2 \cdot \cdot nH_2O$
Magnesian beidellite (β-cerolite)	$Mg_3[Si_4O_{10}][OH]_2 \cdot nH_2O$

The montmorillonites are characterized by the presence of alkaline-earth metals; the ordinary formula, taking these constituents into account, is $m\{(Mg, Ca)_3[Si_4O_{10}][OH]_2\} \cdot p\{(Al, Fe)_2[Si_4O_{10}][OH]_2\} \cdot nH_2O$, where the ratio $m:p$ is generally 0.8-0.9.

According to Hofmann, Endell, Wilm, Marshall, and Hendricks, montmorillonite is a three-layer structure: two outer silica tetrahedral sheets and a central alumina-hydroxyl octahedral sheet. The tetrahedral and octahedral sheets are combined so that the tips of the tetrahedrons of each silica sheet and the tips of the hydroxyl layers of the octrahedral sheet form a common layer (Figs. 25 and 26).

A specific characteristic of the montmorillonite strucutre is that the molecules of water or other polar substances (some organic molecules, for instance) can enter between the unit layers, causing the lattice to expand in the c direction (at right angles to the disposition of the layers). The c-axis dimension is therefore variable, ranging from about 9.6 A when polar molecules are absent between the unit layers to almost complete separation of the individual layers in some cases.

According to R. E. Grim [1956] the exchangeable cations occur between the silicate layers, and the interplanar distance along the c axis of completely dehydrated montmorillonite depends somewhat on the size of the interlayer cation, being larger the larger the cation. In the case of adsorption of polar organic molecules between the silicate layers, the c-axis dimension also varies with the size and geometry of the organic molecule. The thickness of the water layers between the silicate units depends on the nature of the exchangeable cation at a given water-vapor pressure. Under ordinary conditions a montmorillonite with Na^+ as the exchange ion frequently has one molecular water layer between the silicate layers and a c-axis spacing of about 12.5 A. A c-axis spacing of 15.5 A corresponds to montmorillonite with Ca^{++} in which there are generally two molecular layers. The expansion properties are reversible [Grim, 1956, pp. 74-75]. It has been discovered that the thickness of the water

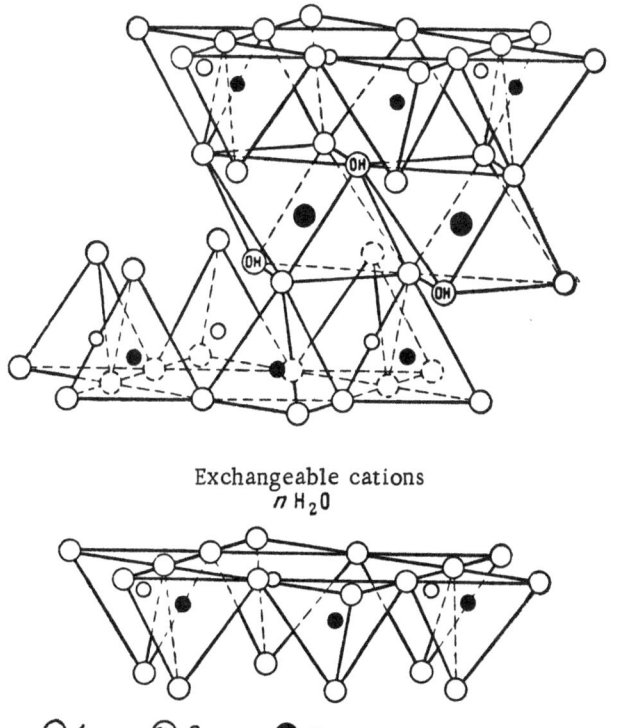

Exchangeable cations
n H$_2$O

O 1 (OH) 2 ● 3
O & ● 4

Fig. 25. Diagrammatic sketch of the structure of montmorillonite. (According to Hofmann, Endell, and Wilm, 1933,
Marshall, 1935; and Hendricks, 1942.) 1) Oxygen, 2) hydroxyl, 3) aluminum, iron, magnesium, 4) silicon, occasionally aluminum.

layers between successive silicate layers is an integral number of molecules; i.e., the water layer has a thickness
of one, two, three, or four molecular layers. A natural montmorillonite may have a regular ordering of a single
thickness of water layers, or it may be a random mixture of water layers of variable thicknes. Roth has shown
that important physical characteristics of montmorillonitic clays are related to the nature of the alteration of
molecular water layers (regular or random) between the silicate layers.

Beidellite. This a clay mineral with the structural formula Al$_2$(AlSi$_3$O$_{10}$)(OH)$_2$ · nH$_2$O or Al$_2$(Si$_4$O$_{10}$)(OH)$_2$ ·
· nH$_2$O, possibly interlayered. The empirical formula of beidellite is Al$_2$O$_3$ · 3-3.5SiO$_2$ · 4H$_2$O. The mineral
contains 20-27.6% Al$_2$O$_3$, 45-50% SiO$_2$, and 0.8% Fe$_2$O$_3$. Some investigators express doubt that beidellite exists
as an independent mineral. According to many investigators it is more likely a combination of "mixed crystals"
with fine-layered structure; i.e., its crystal lattice contains alternations of montmorillonite and kaolinite layers.
The hardness of beidellite is 1.5, the specific gravity 2.2-2.6. The color is light blue, greenish, white, yellowish.
The indices of refraction are Ng = 1.531-1.568, Nm = 1.530-1.564, Np = 1.494-1.523 to 1.548; Ng−Np = 0.003-
0.030.

Ferruginous beidellite, or ferribeidellite, contains up to 18.5% Fe$_2$O$_3$ and more (the iron isomorphously replaces aluminum); in this variety Ng = 1.572, Nm = 1.570, Np = 1.523; Ng−Np = 0.049. The indices of refraction
increase with increase in Fe$_2$O$_3$ content. In the ferribeidellites and ferrimontmorillonites from the Southern Urals
(I. I. Ginzburg), Ng−Np ranges from 0.004-0.01 to 0.020-0.028, rarely up to 0.049. Beidellite, like all minerals
of the montmorillonite group, has a distinct capacity for cation exchange and tends to swell in water. It is one
of the chief minerals in bleaching clays, particularly in bentonites. The thermal curve of beidellite is similar to
that for montmorillonite (see below, p. 61), but it differs in having a noticeable or marked exothermic effect
at 860-925° and in that the second endothermic effect generally occurs at a lower temperature (500-600°).

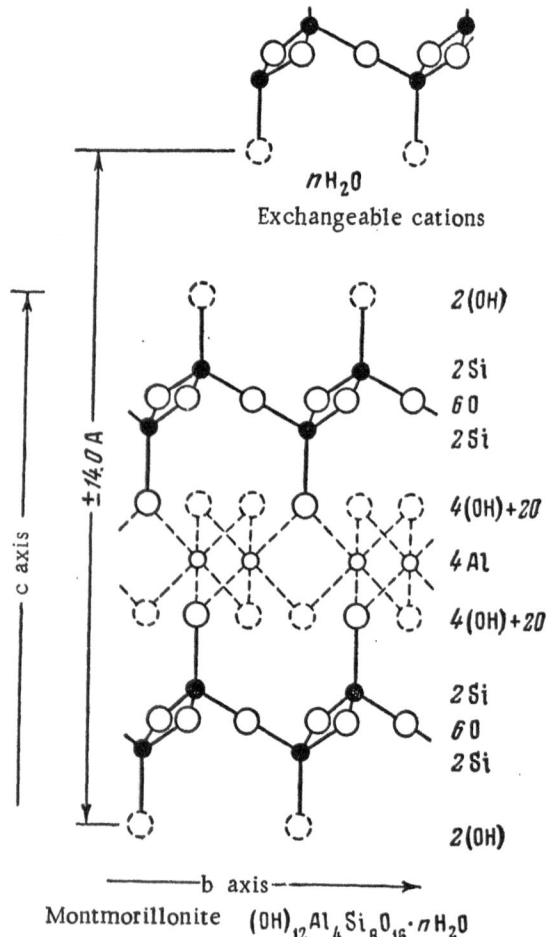

$n H_2O$

Exchangeable cations

2(OH)

2 Si

6 O

2 Si

4(OH)+2O

4 Al

4(OH)+2O

2 Si

6 O

2 Si

2(OH)

c axis

± 14.0 A

b axis

Montmorillonite $(OH)_{12}Al_4Si_8O_{16} \cdot n H_2O$

Fig. 26. Schematic presentation of the structure of montmorillonite. (After Edelman and Favajee, 1940.)

Ross and Hendricks did not distinguish between beidellite and montmorillonite, but combined them into a single group. According to I. I. Ginzburg, the term "beidellite" should be reserved for minerals of the montmorillonite group with an $Al_2O_3 : SiO_2$ ratio ranging from 1 : 3.1 to 1 : 3.5-3 75; minerals with a ratio of 1 : 3.75-1 : 4 and lower should be called montmorillonite; if SiO_2 is greater than 4, there is, according to I. I. Ginzburg, an isomorphous mixture of montmorillonite and β-cerolite.

In beidellite, silicon may be replaced by aluminum (from 1 to 0.75 atom), whereas in montmorillonite such replacement is absent or is but slightly effective.

According to I. I. Ginzburg beidellites may be subdivided into two groups: one with $Al_2O_3 : SiO_2$ ratio about 1 : 3, the other with the ratio 1 : 3.5 or somewhat less (to 1 : 3.75). Minerals of the first type are commonly representatives of pure magnesian (β-cerolite) or magnesium-aluminous varieites; they consequently have less tendency to swell and less capacity for adsorption.

D. P. Serdyuchenko [1956] has classified the montmorillonite minerals on the basis of the composition of the most stable tetrahedral layers in the crystal lattice: 1) montmorillonites in the narrow sense (Si_{iv} = 4.0-3.8), 2) beidellites (Si_{iv} = 3.79-3.60), 3) saponites (Si_{iv} = 3.59-3.40, and 4) parahalloysites or low-silica montmorillonites (Si_{iv} = 3.39 to 3.14-3.00).

I. I. Ginzburg and I. A. Rukavishnikova [1951, p. 592] have observed that beidellite is chiefly formed during weathering of dark minerals (pyroxenes, amphiboles, epidote-group minerals), whereas montmorillonite forms during weathering of basic feldspars.

Beidellitic clays are widespread, commonly in accumulations containing ore deposits. M. A. Rateev, following N. E. Vedeneeva and M. F. Vikulova, recommends the use of organic dyes for rapid discrimination of beidellitic clays, especially to distinguish them from montmorillonitic clays. Beidellitic clays are of marine origin, commonly forming in noncarbonate deposits. M. F. Vikulova [1952] and M. A. Rateev [1954] associate the mineral with the reworking of hydromicaceous material during diagenesis of marine sediments.

Montmorillonite. This is a clay mineral with the structural formula $Al_2(Si_4O_{10})(OH)_2 \cdot nH_2O$; a more common form of this formula, with due consideration to the content of alkaline earths, is $[Al_{1.67}(Mg, Ca)_{0.33}]$ $[Si_4O_{10}][OH]_2 \cdot nH_2O$. The empirical formula is $Al_2O_3 \cdot 4SiO_2 \cdot nH_2O$, or, considering the alkaline-earth content, approximately $m(Mg, Ca)O \cdot Al_2O_3 \cdot 5SiO_2 \cdot H_2O$ 5aq. The chemical composition of montmorillonite fluctuates with changes in the water content: 11-22% Al_2O_3, 48-56% SiO_2, 12-24% H_2O, 4-9% MgO, 0.8-3.5% CaO, and 0-5% Fe_2O_3.

The specific gravity of the mineral is 2.0-2.2. The color is light blue, greenish, green, white, yellowish. The indices of refraction are Ng = 1.502 to 1.522-1.552, Nm = 1.502 to 1.521-1.551, Np = 1.485 to 1.503-1.533; Ng−Np = 0.012-0.034.

Ferruginous montmorillonites, or ferrimontmorillonites, are distinguished by a high content (more than 5%) of Fe_2O_3 (the iron isomorphously replaces aluminum), an increase in Fe_2O_3 content being accompanied by higher refractive indices: Ng = 1.565-1.610, Nm = 1.555-1.600, Np = 1.543-1.589; Ng−Np ≅ 0.02-0.03.

Marked swelling from moisture and base exchange are very characteristic of montmorillonite (in clays). Montmorillonite is the chief mineral in bleaching clays.

Montmorillonites proper (or aluminous montmorillonites), generally having refractive indices lower than that for Canada balsam, may be recognized by cutting down the light with the iris diaphragm, or generally by decreasing the amount of light coming through the thin section; the relief is slightly negative; corresponding zones affect polarized light. More precise determinations, especially for cryptocrystalline varieties, may be made by differential thermal analysis, x-ray measurements, and electron micrography. Montmorillonites proper and beidellites are also distinguished from each other and from the remaining principal groups of clay minerals by dyes — spectrophotometric analysis [Vedeneeva and Vikulova, 1952, 1956].

The thermal curves of montmorillonite is characterized by four reactions: 1) the smallest endothermic effect at 120-200°, due to elimination of weakly held interlayer water; 2) an endothermic effect at 550-700° or 600-730° (the "pyrophyllite" disturbance), marking the expulsion of hydroxyl water; 3) an endothermic effect at 780-880°, reflecting either the decomposition of anhydrous montmorillonite or the elimination of the remaining hydroxyl water; 4) an exothermic effect at a temperature near 950-1000°, due to crystallization of spinel and other minerals, forming from the decomposition products of the montmorillonite.

Rocks consisting chiefly of montmorillonite possess a number of characteristic properties: they are hydrophilic, tend to swell (increasing in volume 8-10 times), are plastic, and have adsorbent capacity Alkalic bentonites swell markedly, are thixotropic (i.e., they go from a liquid suspension to a solid gel on standing, under isothermal conditions), and on stirring they acquire incipient tenacity. Alkaline-earth bentonites by base exchange easily change to alkalic forms, acquiring the properties of the latter; they themselves swell but slightly or not at all.

Montmorillonitic clays are widespread in sedimentary rocks, commonly in sequences containing ore deposits. They form in basins under alkaline or weakly alkaline conditions; in eluvial or slope-wash deposits, ferrimontmorillonites and montmorillonites may be associated with weathering zones on predominantly basic or alkalic igneous rocks. Montmorillonitic and beidellitic clays or montmorillonite-bearing rocks are commonly decomposition (alteration) products of volcanic ashes and tuffs. Individual forms of montmorillonitic clays, having formed by decomposition of ashes and having been studied in certain regions, have received local names: bentonite, keffekilite, nalchikinite, askanite, Fuller's earth (gumbrine), giliabit, and others.

I. D. Sedletskii [1939$_2$] obtained synthetically a mineral rich in sodium, called gedroitzite, under alkaline (strongly alkaline) conditions characteristic of solonets (strongly alkaline soil). The synthesis was continued for 5-7 years, the final formation of the gedroitzite and the typical x-ray powder photograph (complete crystallization) were made at the end of the period. According to I. D. Sedletskii, gedroitzite has the formula $Na_2O \cdot Al_2O_3 \cdot 3SiO_2 \cdot 2H_2O$; not only is sodium present (13.8%) but potassium (7.3%) also. Characteristic diffraction rings have been noted for gedroitzite; a differential thermal curve has been obtained (on the whole, of a mineral in the montmorillonite group); and a dehydration curve has been plotted. The refractive index of gedroitzite is 1.483 ± 0.003.

The cited data indicate gedroitzite is an alkalic (predominantly sodic) analogue of beidellite; clearly there are also analogues of montmorillonite proper.

In nature gedroitzite was first discovered as an impurity in the saline horizon of sodic alkaline soils (solonets). It is concentrated in this horizon chiefly in the size fraction 0.2-2.0 μ.

In another paper I. D. Sedletskii [1939$_3$] has noted that only five elements have been found to be absorbed by soils; these are, in decreasing order: C, Mg, H, Na, and K. The minerals in soil colloids are divided chiefly into three groups: 1) magnesium and calcium aluminosilicates (montmorillonites and beidellite), 2) acid hydrogenous minerals (kaolinite, halloysite, and others), and 3) sodium and calcium minerals (gedroitzite, sericite).

As I. D. Sedletskii noted, cation exchange in podsol soils is 80-96% accomplished by hydrogen and only 20-4% by Ca and Mg, whereas in chernozems the cation exchange is chiefly by Ca and Mg. In solonets (alkaline soils) sodium is the principal agent in the exchange (50 to 70%), Ca and Mg contributing only 30%.

"Clay Group" Magnesian Silicates

These are hydrous magnesian aluminosilicates with $Al_2O_3 : SiO_2 = 1:4$ and hydrous magnesium silicates without aluminum. We shall consider only palygorskite and sepiolite in this subgroup.

Palygorskite (Mountain Leather, Mountain Cork, etc.). This is a magnesian silicate mineral, normally with the structural formula $MgAl_2[Si_4O_{10}][OH]_2 \cdot 4H_2O \cdot nH_2O$, or with the empirical formula $MgO \cdot Al_2O_3 \cdot 4SiO_2 \cdot 5H_2O$. The ratio of Al to Mg is variable: varieties rich in Al_2O_3, with the ratio $Al:Mg \geq 1:1$ palygorskite (α when Al > Mg, and β when Al:Mg = 1:1), whereas varieties poor in Al_2O_3, approaching sepiolite in composition, are called pilolite (α when Mg Al = 2:1, and β when Mg Al = 3:1); β-palygorskite is the most widespread. The system is monoclinic or orthorhombic. The hardness of palygorskite is 2.0-2.5, the specific gravity 2.1-2.3 (increasing with increase in MgO content and decreasing with increase in Al_2O_3 content). The color is white, occasionally yellowish white or gray with tints of yellow or shades of brown. The indices of refraction are Ng = 1.510-1.560, Nm = 1.51-1.55 (approximately); Ng−Np = 0.015-0.025. The mineral is biaxial, negative, and has positive elongation.

All mineral varieties in the palygorskite and sepiolite group are characterized by a fibrous, felty structure, by high porosity, and, because of the latter property, by light weight (very low bulk weight), by virtue of which they will float (when dry) on water. In hot sulfuric acid the minerals decompose, with separation of SiO_2. The differential thermal curve of palygorskite shows three endothermic and one exothermic effects: the first endothermic effect, at 100-120°, is due to the elimination of adsorbed water; 2) the second endothermic effect, at 250-350°, is due to the expulsion of zeolitic water; 3) the third endothermic effect, at 400-500°, results from the loss of hydroxyl water from the lattice of the mineral; 4) the exothermic effect, at 810-910°, is caused by crystallization of the decomposition products.

Palygorskites generally form during weathering of rocks relatively rich in MgO. They occur in nests and irregular tabular deposits in sedimentary rocks. They are frequently found in limestones, dolomites, sand-silt-clay rocks, rarely in veinlets in sedimentary iron ores (derived from serpentinites) and even in partially weathered serpentinites themselves; a mineral resembling α-palygorskite has been found in fossil and relict cavities in alkaline soils (solonets). Palygorskite forms in more or less alkaline environments (from alkaline to almost neutral).

Sepiolite. This is a hydrous silicate of magnesium, some varieties of which have the structural formula $Mg_8(Si_4O_{11}) \cdot H_2O \cdot nH_2O$ and the empirical formula $3MgO \cdot 4SiO_2 \cdot 2H_2O$; other varieties correspond to the formula $2MgO \cdot 3SiO_2 \cdot 2H_2O$. The mineral occurs in fibrous, felty masses.

Chemical analyses of the fraction < 0.001 mm from normal sedimentary sepiolite are shown in Table 15, the data taken from M. A. Rateev and D. D. Kotel'nikov [1956].

Very lightweight material, forming a fibrous, felty mixture with amorphous substance, is called meerschaum. Crystalline (monoclinic) fibrous varieties are generally called α-sepiolite, and colloidal and microcrystalline varieties are called β-sepiolite. The hardness of sepiolite is 2.0-2.5-3.0, and the specific gravity is near 2. The color is white, grayish white, yellowish. The indices of refraction are Ng = 1.520-1.546, Np = 1.496-1.536; Ng−Np = 0.009-0.022; the elongation is positive, the extinction parallel; in isotropic colloidal varieties of sepiolite, N = 1.505-1.517.

Sepiolite forms veinlets, seams, lenses, and nodules, and is found disseminated in other rocks. In the dry state it is dense; when moist it is viscous, waxy, sometimes with a pasty consistency.

Sedimentary sepiolites are characterized by Nm = 1.501-1.510 and by a relative lack of water. A typical differential thermal curve of α-sepiolite shows a very small endothermic effect at 110-150°, due to the elimination of hygroscopic water; a second endothermic effect at 350-430°, due to the expulsion of zeolitic water; third and fourth endothermic effects at 525-620 and 750-830°, caused by the separation of hydroxyl water and by the destruction of the lattice; an exothermic effect at 820-860°, related to the recrystallization of amorphous material, and later (at 1000°) to the formation of enstatite from the decay products of the lattice. Sepiolite is decomposed in hydrochloric acid, with the separation of silica.

Sepiolite is a product of surface weathering [Ginzburg and Rukavishnikova, 1951], an authigenic mineral in association with opal, dolomite, and magnesite, or, finally, a chemical precipitate on the floors of aqueous basins; it forms in an alkaline environment.

Sepiolite is restricted to sedimentary carbonate rocks (limestones, dolomites, marls) or to basic igneous and metamorphic rocks. In sedimentary rocks, generally in dolomites, sepiolite forms individual clay seams.

Sepiolite (α-sepiolite and Mg-silicate type sepiolite) has been found in sedimentary rocks both in Recent and Mesozoic-Cenozoic deposits (for example, in Lower Cretaceous, Callovian, Upper Maikopian, and other formations) and also in Upper Paleozoic sequences. According to M. A. Rateev and D. D. Kotel'nikov [1956, 1957], sepiolite is occasionally found in the Upper and Middle Carboniferous formations of the Russian platform, constituting individual seams of sepiolitic clays in dolomite zones. The mineral is found under similar conditions in the Aleksin strata, and, frequently, in the Tarussa strata of the Moscow Basin. The lowest stratigraphic level

TABLE 15

Chemical Composition of Sepiolite

Components	Sample 2312 Krashaya Polyana	Sample 644, Melekess	Sample 643 Melekess	Sample 2180, Lyskovo
SiO_2	52.68	53.03	48.53	55.97
Al_2O_3	6.19	3.34	3.44	0.20
Fe_2O_3	0.59	1.29	2.78	0.20
FeO	0.48	0.38	0.43	Traces
CaO	—	0.89		—
MgO	20.89	23.43	26.03	24.81
Na_2O	1.39	1.44	1.61	—
K_2O	0.82	—	—	—
H_2O+	6.09	7.01	6.94	8.35
H_2O-	10.87	9.14	10.24	10.47
Totals	100.00	99.95	100.00	100.00

(thus far known) in which sepiolite has been found lies in the Dankov-Lebedyan strata of the Moscow Basin, according to D. D. Kotel'nikov.

M. A. Rateev and D. D. Kotel'nikov have stated that sepiolite is generally associated with dolomite, having formed by chemical sedimentation in an arid climate, during silicification; but Mg-silicates of the palygorskite type form chemically in a normal marine environment, during marked introduction of silica.

Calcium and Sodium Hydrous Aluminosilicates

This group includes authigenic zeolites and analcime, all most frequently encountered in sedimentary rocks.

Zeolites

These are hydrous aluminosilicates, chiefly of Ca and Na, less commonly of K, and partly of Ba and Sr. The general formula for zeolites may be represented by $A_m X_p O_{2p} \cdot nH_2O$, where X = Si and Al, and A represents alkali and alkaline-earth metals; no definite relationship between content of alkali and of silica has been observed [Betekhtin, 1950].

Zeolites are characterized by the fact that on gentle heating they may lose their water gradually without destruction of the crystal lattice as a whole; this water may then be absorbed in a reversible reaction, or it may be partly replaced by molecules of other substances. So-called zeolitic water is distinguished from crystallization water precisely by the fact that during heating it is given off gradually, without any sudden jumps at any particular temperature. Another characteristic property of most zeolites, as noted by A. G. Betekhtin, is the ease with which exchange occurs between the cations neutralizing the charge in the body of the crystal lattice and cations in the surrounding solution. This property is utilized in practice for softening hard water, i.e., as a "permutite."

Zeolites form under low pressures in the latest low-temperature stages of hydrothermal (apomagmatic) processes; they are also widespread in sedimentary formations.

These zeolites are commonly associated with volcanic activity, being the products of alteration of volcanic ash on the floors of basins or of alteration of tuffs; in addition they occur in sedimentary silica-carbonate-clay rocks. The potassium-calcium zeolite phillipsite occurs in Recent abyssal oceanic deposits, particularly in red deep-water clays and in radiolarian oozes; this mineral has a structural formula of $(K_2, Ca) [Al_2Si_4O_{12}] \cdot 4.5H_2O$, and represents the decomposition product of volcanic glass (ash).

Other zeolites are found in sedimentary rocks, locally in very large quantities: mordenite, laumontite, heulandite, natrolite, and, rarely, chabazite.

All sedimentary zeolites have refractive indices lower than that for Canada balsam (from 1.46 to 1.53-1.54) and very low birefringence (from 0.001 to 0.014). The hardness is 3.5-5.5, the specific gravity 2.0-2.78. The color is pure white or white with tints of other colors.

Three subgroups are distinguished among the zeolites:

I. Chabazite subgroup:

Chabazite	$(Ca, Na_2)[AlSi_2O_6] \cdot 6H_2O$, trigonal system,
Laumontite.	$(Ca, Na_2)[AlSi_2O_6] \cdot 4H_2O$, monoclinic system.

II. Natrolite-thomsonite subgroup:

Natrolite	$Na_2[Al_2Si_3O_{10}] \cdot 2H_2O$, orthorhombic system,
Thomsonite	$Ca_2Na [Al_5Si_5O_{20}] \cdot 6H_2O$, orthorhombic system,
Scolecite	$Ca [Al_2Si_3O_{10}] \cdot 3H_2O$, monoclinic system.

III. Heulandite and phillipsite subgroup:

Heulandite	$(Ca, Na_2)[AlSi_3O_8]_2 \cdot 5H_2O$, monoclinic system,
Phillipsite	$(K_2, Ca)[Al_2Si_4O_{12}] \cdot 4.5H_2O$, monoclinic system,
Mordenite	$(Ca, Na_2, K_2)[Al_2Si_9O_{22}] \cdot 6H_2O$, monoclinic system.

Analcime, $Na(AlSi_2O_6) \cdot H_2O$, in recent times has been assigned to the leucite group rather than to the zeolite group.

We shall consider the zeolite subgroups in the order of their distribution in sedimentary formations.

Heulandite and Phillipsite Subgroup

Mordenite — $(Ca, Na_2, K_2) [Al_2Si_9O_{22}] \cdot 6H_2O$, or $(Ca, Na_2, K_2)O \cdot Al_2O_3 \cdot 9SiO_2 \cdot 6H_2O$. The chemical composition of sodium mordenite is 66.81% SiO_2, 12.59% Al_2O_3, 3.45% CaO, 3.83% Na_2O, and 13.32% H_2O. The mineral is in the monoclinic system. The hardness is 3-4 (most commonly near 4), the specific gravity 2.15. Crystals are tabular and prismatic. The color is white with yellowish or rose-colored tints. There is cleavage along (100) and (101). The indices of refraction are: Ng from 1.473 to 1.482-1.489 and more, Nm from 1.472 to 1.480-1.485, Np from 1.471 to 1.478-1.483; Ng−Np = 0.002-0.005; the optic sign may be either positive or negative. The mineral is known to occur in basalts (in amygdules) and in sedimentary rocks, where the calcium-sodium variety predominates — $(Ca, Na_2)(Al_2Si_9O_{22}) \cdot 6H_2O$.

In recent years (1945-1955) N. V. Rengarten, G. V. Gvakhariya, G. I. Bushinskii, V. S. Vasil'ev, and other investigators have established the fact that mordenite is widespread in sedimentary rocks in the USSR, particularly in chalk, chalk-like and other marls, opaline clays, clays, sands, sandstones, and, in part, in phosphorites of Mesozoic and Tertiary age. The refractive indices of sedimentary mordenite generally range from 1.480 to 1.487-1.489, the birefringence being very weak, occasionally absent entirely; the mineral is frequently found in very small prisms (≤ 0.01 mm), sometimes drawn out in acicular form. Apparently some authors previously mistook mordenite in sedimentary deposits for authigenic feldspar.

Sedimentary mordenite is generally restricted to marine deposits (except where eluvium has developed on these deposits) and is a diagenetic and epigenetic mineral; it is found with glauconite, frequently with opal, and occasionally with chalcedony, phosphate minerals, montmorillonite, calcite, FeS_2, chabazite (?), and hydromicas.

Phillipsite — $(K_2, Ca) [Al_2Si_4O_{12}] \cdot 4.5H_2O$. The chemical composition is 44-48% SiO_2, 22-24% Al_2O_3, 4-11% K_2O, 3-8% CaO, 15-17% H_2O; Na_2O is also present (sometimes up to 6%).

The crystal system is monoclinic. The hardness is 4.0-4.5, the specific gravity 2.2; the crystals have a prismatic form, but are commonly twinned. The mineral is colorless, or white with grayish, yellowish, and reddish tones. Cleavage is present along (001) and (010). The indices of refraction are $Ng = 1.500-1.503$, $Nm = 1.497-1.500$, $Np = 1.493-1.498$; $Ng-Np = 0.005$ (approximately); the optic sign is positive, and the extinction angle ($c \wedge Ng$) is 10-30°.

Phillipsite forms characteristic twins or "quadruplets," sometimes cruciform. It dissolves in hydrochloric acid, with the separation of gelatinous or flocculent silica. As noted above, phillipsite occurs in modern deepwater red oceanic clays and radiolarian oozes, where it is the alteration product of volcanic glass, it is also found in cavities in various volcanic rocks (in amygdules and secretions).

Desmine (stilbite) — $(Na_2, Ca) [Al_2Si_6O_{16}] \cdot 6H_2O$. This mineral is monoclinic. The hardness is 3.5, the specific gravity 2.09-2.20. The color is white with yellowish or reddish tones. Cleavage is present along (010) and (100). The indices of refraction are $Ng = 1.500$, $Nm = 1.498$, $Np = 1.493$; $Ng-Np = 0.007$; the mineral is biaxial, negative. The extinction angle ($c \wedge Np$) is 5°. Desmine is found in twins (commonly "quadruplets," sometimes cruciform) or in sheaf-like aggregates of complex, twinned crystals. It decomposes in hydrochloric acid. Desmine (stilbite) is found in cavities and fractures of volcanic rocks; it is also encountered in hydrothermal ore veins and in normal sedimentary sandstones [Ermolova, 1953, 1956].

Epistilbite (epidesmine), which formed during diagenesis of sediments, has been described from analcime-bearing sedimentary rocks in the Carboniferous sequence of Tuva [Bur'yanova, 1954]. This mineral is orthorhombic and is identical in composition to desmine; $Ng \cong 1.517$, $N = \cong 1.506$; $Ng-Np = 0.011$; the optic sign is positive, and c $Ng = 9-10°$.

Heulandite — $(Ca, Na_2) [AlSi_3O_8]_2 \cdot 5H_2O$. This is a monoclinic zeolite. The hardness is 3.5-4, the specific gravity 2.18-2.22. Crystals have a tabular appearance. The color is white or colorless, rarely yellow. The indices of refraction are $Ng = 1.496-1.505$, $Np = 1.488-1.498$; $Ng-Np = 0.007-0.008$; the optic sign is positive. Heulandite is found in single crystals or in lamellar masses with parallel intergrown plates. It decomposes easily in hydrochloric acid, with the separation of gelatinous silica.

Heulandite is found in cavities (amygdules) in volcanic rocks, in ore veins, and as an authigenic mineral in sedimentary rocks, sandstone, for instance [Gilbert and McAndrews, 1948; Kossovskaya, 1954]. It forms during diagenesis and epigenesis by alteration of clastic material in sandy sedimentary deposits.

Chabazite Subgroup

Laumontite — $(Ca, Na_2) [AlSi_2O_6] \cdot 4H_2O$. The formula is approximate.

The mineral is monoclinic. The hardness is 3-4, the specific gravity 2.23-2.41. Crystals are prismatic or equant. The color is white, with various tints, the mineral is colorless; minerals are occasionally brick red from fine inclusions of hematite. Cleavage occurs in three directions: (010), (110), and (100). The indices of refraction are $Ng = 1.510-1.529$, $Np = 1.499-1.514$; $Ng-Np = 0.011-0.016$, generally about 0.012, the mineral is optically negative. Extinction is inclined: $c \wedge Ng = 20-40°$. Laumontite occurs in individual crystal, in aggregates of crystals, or in the cement of sandstones. In hydrochloric acid it decomposes and yields gelatinous silica.

Laumontite is found in cavities in volcanic rocks, in ore deposits, and, in noticeable quantities, in the cement of sandstones. It has been described by N. V. Rengarten [1950] from the Lower Jurassic rocks of the Northern Caucasus, and by E. Z. Bur'yanova [1954, 1956] from Lower Carboniferous sedimentary rocks and from Middle Devonian sandstones in Tuva; A. G. Kossovskaya and V. D. Shutov [1956] have also described the mineral from Mesozoic rocks in Yakutia.

N. V. Rengarten believes that laumontite is a primary component in the sedimentary rocks he examined; he has established the following sequence of formation of authigenic minerals in the Lower Jurassic ferruginous sandstones: laumontite, chlorite, analcime.

65

E. Z. Bur'yanova thinks that the laumontite cement in the Middle Devonian sandstones of Tuva was formed by crystallization of an alumina-silica gel which had absorbed Ca^{++} and Na^+ during early diagenesis. The cement is these sandstones consists of irregular, alternating sections of calcite and laumontite.

A. G. Kossovskaya and V. D. Shutov have described laumontite as a product of epigenesis in arkosic sandstones in the Mesozoic sequence of Yakutia.

Chabazite — $(Ca, Na_2)[AlSi_2O_6] \cdot 6H_2O$. This mineral is in a group of zeolites belonging to the trigonal system. The hardness is 4-5, the specific gravity 2.08-2.16. Crystals are rhombohedral, nearly cubic. The color is white, with reddish or brownish tones; the mineral is sometimes colorless. Rhombohedral cleavage is distinct. The indices of refraction are Ne (or Ng) = 1.480-1.490, No (or Np) from 1.474-1.478 to 1.485; Ne—No (or Ng—Np) ranges from 0.002 to 0.005-0.014; the mineral is uniaxial (most frequently positive) or biaxial (positive or negative) with a 2V ranging up to 32°. It is found in druses, incrustations, secretions, and dense aggregates. It decomposes in hydrochloric acid, with the separation of silica.

Laumontite is frequently found in amygdules in volcanic rock; it occurs in fossil mollusc shells (Iceland) and as cement in sandstones; it forms during epigenesis and weathering of bauxites, and is sometimes present in phosphorites [Bushinskii, 1954].

Natrolite Subgroup

Natrolite — $Na_2[Al_2Si_3O_{10}] \cdot 2H_2O$. This is an orthorhombic zeolite. The hardness is 5.0-5.5, the specific gravity 2.2-2.5. Crystals are prismatic. The color is white, with yellowish, greenish, and reddish tones, or the mineral is colorless. Cleavage is present along (110). The indices of refraction are Ng = 1.485-1.493, Nm = 1.476-1.482, Np = 1.473-1.480; Ng—Np = 0.011-0.013; the mineral is optically positive and shows distinct negative relief. It is frequently found in radiating aggregates, normal fibrous incrustations, and spherulites and fibrous masses. Natrolite decomposes easily in hydrochloric acid.

It occurs in amygdules and geodes of volcanic rocks, as a late hydrothermal mineral in some pegmatites, and in small quantities in sedimentary rocks (sandstones and others).

* * *

Analcime — $Na[AlSi_2O_6] \cdot H_2O$, or $Na_2O \cdot Al_2O_3 \cdot 4SiO_2 \cdot 2H_2O$. This mineral is isometric. The hardness is 5.0-5.5, the specific gravity 2.2-2.3. It is colorless or white with grayish, greenish, or reddish tones. It is generally isotropic, with N about 1.487, but it may be biaxial, negative, with Ng = 1.487 and Ng—Np = 0.001; in general N may range from 1.489 to 1.478. Weak birefringence has been detected in free crystals, a phenomenon probably associated with loss of water.

Analcime is characterized by very low indices of refraction and by optic isotropy (but with crystalline structure); rarely, weak birefringence may be observed. The mineral may be easily distinguished from volcanic glass by its lower refractive index, even when crystal boundaries are absent; on the other hand, analcime has a higher index of refraction than opal. In hydrochloric acid analcime gelatinizes (decomposes), yielding a slimy precipitate of silica.

Analcime was previously known as a product of low-temperature hydrothermal activity, developing during post-volcanic processes at temperatures commonly below 100°. At present it is also widely recognized as an authigenic mineral of sedimentary origin, forming during exogenetic processes: by alteration of volcanic glass on the floors of basins or through the action of circulating waters in rocks; in muds by precipitation of hydrogels of SiO_2 and Al_2O_3, with absorption of Na ions, in recent sediments; by precipitation from circulating solutions during epigenesis; and, finally, in connection with soil-forming processes.

Analcime was described by Vl. Malyshek [1936] from sands in the red-bed sequence in the region of Neftedag and from rocks of the productive beds of the oil-producing sequence on the Apsheron Peninsula; by P. P. Avdusin [1938] as crystals and cement in sandstones in the Jurassic formations of the Ural-Emba region; by N. V. Rengarten from sandstones of the Kazan series in the Kirov region [1940] and in Lower Jurassic rocks of the Northern Caucasus [1950]. In Georgia, analcime has been recognized by G. S. Dzotsenidze [1943], G. V. Gvakhariya [1951], and E. P. Ermolova, [1952]. In addition, analcime has been described by A. M. Boldyreva [1953] and other investigators.

Feldspars are among the most widespread of rock-forming minerals in igneous and metamorphic rocks, and are also among the most abundant allogenic (fragmental) rock-forming components of sedimentary rocks. They have been repeatedly described in the literature.

Authigenic feldspars in sedimentary rocks are by no means found everywhere; they generally occur in small quantities, but in places they are abundant. The most abundant types found in sedimentary rocks (sandstones, dolomites, and limestones) are authigenic albite, microcline, orthoclase, and less abundantly, acid plagioclase.

Chemically the feldspars represent isomorphous mixtures of the following aluminum-silicon salts of potassium, sodium, and calcium: $KAlSi_3O_8$, $NaAlSi_3O_8$, and $CaAl_2Si_2O_8$. The hardness is 6.0-6.5, the specific gravity 2.5-2.7 (3.4 for the rare barium feldspars). Feldspars crystallize in the monoclinic and triclinic systems.

The feldspars are divided into three subgroups: 1) plagioclases, or soda-lime feldspars; 2) alkali feldspars (potassium-sodium feldspars); 3) rarely encountered potassium-barium feldspars, represented by isomorphous mixtures of the orthoclase molecule ($KAlSi_3O_8$) and the celsian molecule ($BaAl_2Si_2O_8$).

Plagioclases

These are minerals in a group of feldspars that represent an isomorphous mixture of the albite (Ab) molecule ($NaAlSi_3O_8$) and the anorthite (An) molecule ($CaAl_2Si_2O_8$). They crystallize in the triclinic system. The hardness is 6.0-6.5, the specific gravity 2.61-2.77. Crystals are tabular and tabular-prismatic.

According to Tschermak's classification, the molecular composition of the principal types of plagioclases may be represented in the following form:

Albite	$Ab_1An_0 - Ab_6An_1$
Oligoclase	$Ab_6An_1 - Ab_3An_1$
Andesine	$Ab_3An_1 - Ab_1An_1$
Labradorite	$Ab_1An_1 - Ab_1An_3$
Bytownite	$Ab_1An_3 - Ab_1An_6$
Anorthite	$Ab_1An_6 - Ab_0An_1$

However, a more convenient classification of the plagioclases is based on the percentage content of anorthite (An) in the isomorphous mixture:

Albite	(0-10 An)
Oligoclase	(10-30 An)
Andesine	(30-50 An)
Labradorite	(50-70 An)
Bytownite	(70-90 An)
Anorthite	(90-100 An)

Intermediate varieties may be called albite-oligoclase, oligoclase-andesine, etc. Plagioclases are also designated by a number (for example, No. 45) according to the number of anorthite molecules per 100 molecules of plagioclase. The specific gravity increases from albite (2.61) to anorthite (2.77); the indices of refraction increase in the same direction. For albite Ng = 1.536-1.539, Nm = 1.529-1.533, Np = 1.525-1.529; Ng−Np = 0.010-0.011; for anorthite Ng = 1.588, Nm = 1.583, Np = 1.575; Ng−Np = 0.013. Inclined extinction varies in grains on (001) from 4° in albite to 37° in anorthite, and in grains on (010), from 18° in albite to 36° in anorthite. It is very characteristic for plagioclases to show parallel striations of polysynthetic twinning according to the albite law (when one set of bands goes extinct, the other set becomes light, and vice versa). In some specimens perpendicular bands (to the albite twins) occur, representing twinning according to the pericline law; the combination produces a distinct lattice structure that may be distinguished from a similar feature in microcline by straight, or almost straight, edges of the bands. It is significant that the indices of refraction for albite and oligoclase are lower than that for Canada balsam, whereas the indices of the other plagioclases are higher.

In sedimentary rocks, plagioclase occurs both in fragmental grains (normally more or less turbid because of surface alteration) and in authigenic forms (especially albite and oligoclase).

Albite — Na[AlSi$_3$O$_8$], or Na$_2$O · Al$_2$O$_3$ · 6SiO$_2$. Albite is an end number of the plagioclase series, the content of the anorthite molecule ranging from 0 to 10%. The chemical composition of albite No. 0 is 10.79% Na$_2$O, 19.40% Al$_2$O$_3$, and 68.81% SiO$_2$.

The mineral is triclinic. The hardness is 6.0-6.5, the specific gravity 2.61-2.63. The color is most frequently white.

For pure albite the refractive indices are Ng = 1.536, Nm = 1.529, Np = 1.525; Ng−Np = 0.011; the mineral is biaxial and positive. For albite No. 5, Ng = 1.539, Nm = 1.532, Np = 1.529; Ng−Np = 0.010.

Albite is found in acidic magmatic rocks, less abundantly in pegmatites, but very commonly in crystalline schists as a product of regional metamorphism; it forms with other plagioclases in zones of contact metamorphism. Albite is known as an authigenic mineral in sedimentary rocks.

Oligoclase. This name applies to an isomorphous mixture of 90-70% albite (NaAlSi$_3$O$_8$), and 10-30% anorthite (CaAl$_2$, Si$_2$O$_8$) or of Ab$_6$An$_1$-Ab$_3$An$_1$; oligoclase is triclinic.

For oligoclase No. 15, the indices of refraction are Ng = 1.544, Nm = 1.540, Np = 1.536; Ng−Np = 0.008. For oligoclase No. 20 the specific gravity is 2.64, and the indices of refraction are Ng = 1.546, Nm = 1.543, Np = 1.539; Ng−Np = 0.0075; the mineral is biaxial, negative. The refractive indices for oligoclase No. 25 are Ng = 1.549, Nm = 1.546, Np = 1.542; Ng−Np = 0.007. It is thus seen that all indices of refraction increase with increase in percentage of the anorthite molecule. Transitions to albite may be called albite-oligoclase, and transitions to andesine, oligoclase-andesine.

Andesine. This is plagioclase consisting of 70-50% albite (NaAlSi$_3$O$_8$) and 30-50% anorthite (CaAl$_2$Si$_2$O$_8$). The system is triclinic. Andesine No. 40 has a specific gravity of 2.68 and the following indices of refraction: Ng = 1.557, Nm = 1.553, Np = 1.550; Ng−Np = 0.007; the mineral is biaxial, positive. This is the normal feldspar in intermediate or neutral igneous rock.

Labradorite. This is plagioclase consisting of 50-70% anorthite (CaAl$_2$Si$_2$O$_8$) and 50-30% albite (NaAlSi$_3$O$_8$); the system is triclinic. For labradorite No. 52, the specific gravity is 2.69, and the indices of refraction are Ng = 1.563, Nm = 1.558, Np = 1.555; Ng−Np = 0.008; the mineral is biaxial, positive. For labradorite No. 65, Ng = 1.570, Nm = 1.565, Np = 1.561; Ng−Np = 0.009.

Bytownite. This mineral from the plagioclase subgroup is an isomorphous mixture of 70-90% anorthite (CaAl$_2$Si$_2$O$_8$) and 30-10% albite (NaAlSi$_3$O$_8$); the system is triclinic. Bytownite No. 75 has a specific gravity of 2.73, and the indices of refraction are Ng = 1.574, Nm = 1.570, Np = 1.565; Ng−Np = 0.009; the mineral is biaxial, negative. For bytownite No. 85, Ng = 1.578, Nm = 1.574, Np = 1.568; Ng−Np = 0.010. Bytownite is an ordinary mineral in basic igneous rocks.

Anorthite. This is a mineral in the plagioclase subgroup, consisting of 90-100% anorthite (CaAl$_2$Si$_2$O$_8$) and 10-0% albite (NaAlSi$_3$O$_8$); the crystal system is triclinic. The hardness is 6.0-6.5, the specific gravity 2.74-2.76. The color is white, grayish, reddish. For anorthite No. 95, Ng = 1.585, Nm = 1.580, Np = 1.573; Ng−Np = 0.012. The indices of refraction of pure anorthite are Ng = 1.588, Nm = 1.583, Np = 1.575; Ng−Np = 0.013; the mineral is biaxial and negative. Anorthite is a normal mineral in basic igneous rocks. There are single indications that anorthite is present as an authigenic mineral in sedimentary rocks.

Potassium-Sodium Feldspars

In addition to pure potassium varieties — orthoclase, microcline, and sanadine,* there are sodium-bearing varieties of these minerals.

Orthoclase — K[AlSi$_3$O$_8$], or K$_2$O · Al$_2$O$_3$ · 6SiO$_2$. The mineral is monoclinic. The hardness is 6.0-6.5, the specific gravity 2.5-2.6. Crystals are normally prismatic. The indices of refraction are Ng = 1.526 (in general 1.525-1.527), Nm = 1.524 (in general 1.523-1.525), and Np = 1.519 (in general 1.518-1.520); Ng−Np = 0.007; the mineral is biaxial, negative, with (—) 2V = 60-80°. Plates on (001) show parallel extinction; plates on (010) have an extinction angle of 5°.

---

* We shall not discuss high-temperature sanadine and natrosanadine (the 2V in sanadine is small).

Orthoclase is widespread in acid igneous rocks, to some extent in intermediate igneous rocks; it is most frequently found in paleotypal acid volcanic rocks; it is rarer in pegmatites than microcline. It occurs in sedimentary rocks as fragmental grains and as an authigenic mineral.

Microcline — $K[AlSi_3O_8]$, or $K_2O \cdot Al_2O_3 \cdot 6SiO_2$. In composition this mineral is identical to the low-temperature orthoclase and the high-temperature sanadine, but it differs from these monoclinic potassium feldspars in belonging to the triclinic system. The hardness is 6.0-6.5, the specific gravity 2.54-2.57. Crystals generally have a prismatic aspect. The indices of refraction are $Ng = 1.527$ (in general 1.525-1.530), $Nm = 1.525$ (in general 1.523-1.526), $Np = 1.520$ (in general 1.518-1.522); $Ng-Np = 0.007-0.008$; the mineral is biaxial and negative, with (—) 2V near 83°. The indices of refraction and the birefringence of microcline are the same as in orthoclase, from which it is generally easily distinguished by the characteristic grid structure [plates on (001) show an inclined extinction of 16° with the trace of the (010) cleavage].

Microcline is found in intrusive acid and alkalic rocks (granite, granodiorite, syenite, and others). It is also the principal mineral in pegmatite veins and occurs in several metamorphic rocks. In sedimentary rocks its presence is due chiefly to the erosion of acid magmatic and metamorphic rocks. It is also known in sedimentary rocks as an authigenic mineral: in Devonian sands along the Lovat River, in Devonian sandstones in Timan, in dolomites in the environs of Povenets, and in Triassic shell limestones of Alsace and Lotharingia.

Anorthoclase — $(K, Na) [AlSi_3O_8]$. In anorthoclase the content of K_2O exceeds the content of Na_2O; CaO is commonly present as an impurity (at times up to several percent). Anorthoclase crystallizes in the triclinic system; it has a moderate or small optic angle. The specific gravity is 2.56-2.60. The indices of refraction of potassic anorthoclase (that is, anorthoclase rich in potassium) are $Ng = 1.5275$, $Nm = 1.524$, and $Np = 1.520$; $Ng-Np = 0.007-0.008$. Sodic anorthoclase, containing a significant content of $NaAlSi_3O_8$, has the indices $Ng = 1.529-1.530$, $Nm = 1.528-1.529$, and $Np \cong 1.523$; $Ng-Np = 0.006-0.007$. Anorthoclase is biaxial, optically negative, with a 2V ranging from —40 to —60°.

The mineral occurs in lavas and, in general, in sodium-rich magmatic rocks. It is occasionally found in sedimentary rocks, both in clastic grains and as an authigenic mineral.

* * *

Microcline is generally distinguished from orthoclase (in corresponding sections) by its grid structure, but it should not be forgotten that microcline without the grid pattern is frequently found in rocks. Plagioclase is recognized by comparing the indices of refraction with those of quartz, by the extinction angle measured against cleavage planes, and by the extinction angle in a section perpendicular to both the (010) and (001) planes; this latter section has symmetrical extinction relative to the twinning trace, and cleavage along (001) shows distinctly almost at right angles to the trace of the composition plane. (It is desirable to check that the thin section is perpendicular to the cleavage.) An accurate identification of potassium and other feldspars, especially plagioclase, in igneous as well as in sedimentary rocks (in which they may be both authigenic and fragmental), is best made by means of the Federov stage.

It should be noted that previously several authors apparently mistook mordenite (zeolite) in sedimentary deposits (chalks and other rocks) for authigenic feldspar. Moreover, one should naturally not consider idiomorphic feldspars found in tuffs, some tuffaceous sandstones, and other rocks with admixtures of ashy material as authigenic minerals. Finally, some secondary feldspars develop in sedimentary rocks during the stage of incipient (deep regional) metamorphism.

O. M. Ansheles and V. B. Tatarskii [1931] described enlargements of feldspars in Devonian sands along the Lovat River (Leningrad oblast). They found that approximately 65% of the feldspar grains consist of inner fragmental turbid cores, composed of microcline or orthoclase, and outer clear zones with well-defined boundaries, also composed of microcline or orthoclase.

Authigenic alkali feldspars have been described from Oligocene and Miocene deposits in Georgia by E. P. Ermolova [1952, 1956]. A. I. Lebedeva [1956], in studying the Cambrian sands and sandstones of the Leningrad oblast and the Esthonian SSR, found authigenic feldspars to be extensive. The sands and sandstones that have been studied are commonly ferruginous, the cement in the sandstones consisting of calcite and dolomite. Feldspars in the sands generally constitute 5-16%, quartz 83-95% of the rock; mica, carbonates, glauconite, and gypsum are also present. The feldspars are chiefly orthoclase, then microcline, rarely acid plagioclase, most of the grains being rounded and overgrown. A. I. Lebedeva distinguished four varieties of overgrown feldspar: 1) well-defined crystals of orthoclase with hexagonal outline; 2) rims of secondary feldspar (orthoclase, rarely albite) around clastic grains of orthoclase; 3) rims of secondary feldspar (in places orthoclase, in places acid plagioclase) around clastic grains of microcline; 4) rims of feldspar (orthoclase) around clastic grains of plagioclase (albite). A. I. Lebedeva believes that the authigenic feldspars in the Cambrian sands grew in an environment far removed from metamorphism, rather one characteristic of the surficial zones of the sedimentary shell.

Feldspars that occur in limestones are probably partly of hydrothermal origin. V. P. Baturin [1928] showed that there is a relationship between albitization and secondary silicification (quartz development), which are apparently hydrothermal processes, in limestones along the Georgian military road. Discoveries of secondary feldspars in the Kiev chalk [Chirvinskii, 1916] are doubtful, as already noted by V. P. Baturin, because the feldspars from this deposit were distinguished exclusively by means of extinction angles. G. I. Bushinskii [1950, 1954], having recognized the extensive development of mordenite and the absence of authigenic "feldspars" (such as described from writing chalk by P. A. Zemyatchenskii [1916] and V. N. Chirvinskii [1916]) in Cretaceous deposits of the Central chernozem belt, the Ukraine, and other regions, shows that the indicated authors mistook mordenite, in places, for feldspars, elsewhere, for opal. These facts do not preclude the possibility, however, that authigenic feldspars may be found in rocks of Cretaceous age.

Note should also be taken of the work of P. A. Borisov, who described crystals of feldspar and mica from Proterozoic dolomites; the origin of these minerals was apparently related to strong regional metamorphism.

Thus, not only is weathering of feldspars widespread in the surficial zones of the lithosphere, but, under certain conditions, the reverse process — authigenesis — is also effective. As L. V. Pustovalov has emphasized [1956$_2$], secondary feldspars are rather widespread in sedimentary rocks in a number of regions.

Sedimentary Silicates of Copper

Chrysocolla — $CuSiO_3 \cdot nH_2O$ (where n is near 2). The crystal system of this hydrous copper silicate has not been determined; the mineral is generally a typical colloform growth. The chemical composition is variable, and, consequently, the following varieties are distinguished in addition to chrysocolla proper: a) asperolite — a variety rich in water, containing near $3H_2O$; 2) bisbeeite — a variety poor in water, containing one molecule of H_2O; and 3) pilarite — an aluminous variety (up to 17% Al_2O_3). In addition, there are varieties containing up to 7% Fe_2O_3 or up to 7-9% P_2O_5.

The hardness of chrysocolla is about 2, sometimes as much as 4; the mineral is brittle and has a specific gravity of 2.0-2.3. The color is light blue, bluish green, or dark blue; if iron-hydroxide impurities are present, the color is brown; other impurities may make it black. The normal variety (without pigmenting impurities) has a greenish white streak. Chrysocolla is found in opaline masses, in incrustations with a sintered (in places vesicular) surface; i.e., it occurs in colloform growths and in the form of earthy friable masses. Some varieties show a fine crystalline structure in thin section; A. G. Betekhtin [1950] considers this phenomenon to be optical anisotropy due to strain in the colloidal mineral. Different indices of refraction are ascribed to such varieties: Nm = 1.46 or 1.58-1.60 (with Ng−Np ranging from 0.02 to 0.08). The discovery of numerous fibrous incrustations (the fibers normal to the surface) in sandstones (see p. 71) leads one to believe that such material consists of a crystalline mineral of indeterminate crystal system (optically biaxial).

F. V. Chukhrov and F. Ya. Anosov have shown that chrysocolla, in all its varieties, may be considered copper montmorillonite with the following formula $Cu_3(OH)_2(Al_y Si_{4-y}) O_9OH \cdot nH_2O$. This point of view is based on a number of facts: a) similar x-ray powder patterns; b) similar optical properties (they are all biaxial, negative); c) the water, and in part the magnesium and calcium, has an absorbed character; d) minerals of intermediate composition have been discovered; and e) the grains have a lamellar structure.

Apart from the color, the colloform structure, and the low hardness, chrysocolla is characterized by 1) the loss of part of the water during heating below 110°, 2) green coloration of the blow-pipe flame, 3) decomposition in acids, with the separation of powdery silica.

Chrysocolla is a typical supergene mineral, especially characteristic of oxidation zones in copper deposits that are found predominantly in regions of hot, arid climate. Chrysocolla is found with various oxide combinations of copper, forming from solutions in a neutral environment.

L. M. Miropol'skii, when investigating the Permian sedimentary copper ores of the Tatar ASSR [1938, p. 96], noted that chrysocolla plays a "very important role" in that region among other copper compounds and that he had discovered it in a great number of deposits. Chrysocolla occurs most abundantly in fine, fibrous aggregates with parallel-fibrous structure, covering clastic grains (as incrustations) in sandstones. In addition, L. M. Miropol'-skii noted the discovery of fine, fibrous chrysocolla in some cases with radiating fibers and in others with matted fibers. He indicated the following optical properties for the chrysocolla fibers: positive elongation, parallel extinction, $Ng = 1.599$, $Nm = 1.596$, $N= = 1.583$ (?), $Ng-Np = 0.020$, and weak pleochroism (from apple green to almost colorless). L. M. Miropol'skii considered chrysocolla to be a supergene mineral, occurring in bands, sometimes, on malachite, replacing cement in sandstones and frequently forming incrustations, alternating successively with incrustations of malachite.

F. V. Chukhrov and D. G. Sapozhnikov, recognizing sulfide and oxide sandstone copper ores in Dzhezkazgan, have pointed out that chrysocolla is a very widespread mineral in oxidized copper ores [Sapozhnikov, 1948, p. 90]. According to F. V. Chukhrov, chrysocolla occurs in the cement of the sandstones and also in bands at this locality. He associates the formation of chrysocolla with the high content of feldspars in the sandstones, the feldspars yielding free, easily soluble silica during weathering.

4. OXIDE AND HYDROXIDE GROUP

Here we shall consider, apart from the previously described silica group (see pp. 17-21), titanium oxides, iron oxides and hydroxides, aluminum and magnesium hydroxides, manganese oxides and hydroxides, and copper oxides.

Authigenic Sedimentary Titanium Oxides

Titanium oxides in the form of extremely fine-grained rutile and anatase form from ilmenite in igneous, metamorphic, and sedimentary rocks in the surficial parts of the earth; these products of alteration of ilmenite are generally described in the literature as leucoxene. This latter may also form by hydrothermal alteration of titanium-bearing minerals (ilmenite and sphene) in igneous rocks.

Ilmenite (titaniferous iron ore) — $FeTiO_3$, or $FeO \cdot TiO_2$. An opaque ore mineral. Its chemical composition is 36.8% Fe and 31.6% Ti. It crystallizes in the trigonal system. The hardness is 5-6, the specific gravity 4.5-5.0; the mineral is weakly magnetic. The color is iron black or steel black, the streak generally black, and the luster submetallic. The mineral is found in sedimentary rocks in fragmental grains, commonly altered to some degree to leucoxene (extremely fine-grained rutile and anatase), rarely yielding distinct authigenic anatase and rutile.

Rutile — TiO_2. The chemical composition of this mineral is 60% Ti and 40% O. The crystal system is tetragonal. The hardness is 6.0-6.5, the specific gravity 4.2-4.3. Grains (authigenic) are prismatic, columnar to acicular. The color is dark yellow, brown, reddish brown, or red. The indices of refraction are very high: $Ne = 2.903$, $No = 2.616$; $Ne-No = 0.287$; the mineral is uniaxial, positive and the length is positive; red and brown varieties are generally weakly pleochroic; extinction is parallel.

Rutile is characterized by tetragonal crystals of prismatic aspect, sometimes showing elbow twins; by very high refractive indices, manifested by the sharp relief of the grains; and by large birefringence; and by other optical properties. In particular, the very large birefringence means that grains of rutile show no interference colors under crossed nicols. The mineral does not dissolve in acids.

Rutile generally occurs in fragmental grains in sedimentary rocks, but it commonly forms in place during decomposition of ilmenite and, rarely, during alteration of other titanium-bearing minerals; it is also sometimes found in bauxite deposits.

Anatase — TiO_2. This mineral belongs in the tetragonal system. The hardness is 5.5-6.0, the specific gravity 3.80-3.95. The color is brown, yellow, sometimes black. The indices of refraction are very high: $No = 2.550-2.562$, $Ne = 2.488-2.490$; $No-Ne = 0.06-0.074$; the mineral is uniaxial, negative, and pleochroic only in thick sections. In sedimentary rocks anatase generally occurs in tabular and pyramidal grains, sometimes showing parallel striations; fine-grained aggregates are common.

Anatase is distinguished from rutile by a lower specific gravity, by smaller birefringence, commonly by the form of the crystals (especially in sedimentary rocks), appearing as tetragonal plates (square or rectangular sections) and as acute bipyramids. Like rutile, anatase does not dissolve in acids.

It is common to find both fragmental and authigenic grains of anatase in the same sedimentary rock. Most frequently one may find secondary anatase that has formed from the decomposition of fragmental titanium-bearing minerals (ilmenite and others). In bauxites, which consist, as is well known, chiefly of aluminum hydroxides, the content of TiO_2 may reach several percent; such bauxites may be termed titaniferous, in contrast to titanium-free bauxites. In titaniferous bauxites the TiO_2 is found chiefly as fine-grained anatase, which has apparently formed by dehydration of primary titanium dioxide hydrate.

Brookite — TiO_2. This mineral is orthorhombic. The hardness is 5.5-6.0, the specific gravity 3.87-4.08. The color is brown or yellow of various tones. The indices of refraction are very high: Ng = 2.741, Nm = 2.586, Np = 2.583; Ng—Np = 0.158; the mineral is biaxial, positive; as a rule it is not pleochroic or is but slightly pleochroic. The optical properties are frequently anomalous. The grains are elongate-tabular, commonly with striations parallel to the principal axis. Brookite is generally fragmental in sedimentary rocks; sometimes authigenic forms occur.

Recently E. F. Ziv [1956] has traced all stages of alteration of ilmenite in a supergene environment, particularly during weathering in a moist climate. This work has permitted us to refine our views on the secondary products of alteration of ilmentie, which are known in the literature as "leucoxene." During supergene alteration of ilmenite, the mineral gradually changes to extremely fine-grained rutile, which differs from the magmatic form by lower specific gravity (4.11 to 3.96-3.82), by solubility in acids (partially in sulfuric acid and completely in hydrofluoric acid), and, in part, by the birefringence. The differences between secondary supergene rutile and the ordinary variety are explained by the highly disperse state of the former.

According to E. F. Ziv [1956], the supergene alteration of ilmenite follows the following pattern:

Ilmenite
↓
Dull, slightly altered ilmenite
↓
Dark brown, rutilized ilmenite
↓
Light brown, strongly rutilized ilmenite
↓
Secondary rutile, yellowish white, gray
↓
Secondary anatase

The author of the paper disputes the conclusion of M. G. Dyadchenko and A. Ya. Khatuntseva [1954] that the authigenic mineral forming on ilmenite represents an independent mineral, leucoxene, having the variable formula $mTiO_2 \cdot Fe_2O_3 \cdot nH_2O$.

Aluminum Hydroxides

These include the following minerals known to occur in sedimentary rocks: alpha-kliachite (sporogelite), diaspore, boehmite, hydrargillite (or gibbsite).

Alpha-Kliachite (sporogelite or alumogel) — $Al_2O_3 \cdot H_2O$. This is a colloidal amorphous variety (or gel) of diaspore. It has also been called diasporogelite. It is commonly found in small (from 0.001 to 0.01-0.03 mm) round isotropic drop-like bodies cemented together. The index of refraction ranges from 1.57 to 1.68; the value increases with decrease of silica and with increase of alumina. The content of SiO_2 in the form of kaolinite, halloysite, and, especially, opal is clearly shown by the index of refraction (N) of alpha-kliachite: with 2.53% SiO_2, N = 1.68; with 10.21% SiO_2, N = 1.62; and so forth. Sporogelite is distinguished from opal by considerably higher index of refraction (clearly higher than that for Canada balsam) and from optically amorphous calcium phosphates by a negative reaction for phosphate (see p. 85).

Alpha-kliachite is present in bauxites, laterites, and red earths, representing the colloidal monohydrate of alumina. Some authors consider the mineral to be a colloidal variety of boehmite.

Diaspore — $HAlO_2$, or $Al_2O_3 \cdot H_2O$. This mineral is a monohydrate of alumina with the chemical composition of 85% Al_2O_3 and 15% H_2O; it is orthorhombic. The crystal structure is similar to goethite, $HFeO_2$. The hardness is 6-7, the specific gravity 3.3-3.5. The color is whitish, grayish white, yellowish, greenish gray, light brown, and light purple. The indices of refraction are Ng = 1.750, Nm = 1.722, and Np = 1.702; Ng−Np = 0.048; the mineral is biaxial and positive, sometimes showing pleochroism (when Mn or $Fe^{\cdot\cdot\cdot}$ are present).

Diaspore is distinguished from boehmite and hydrargillite by greater birefringence (bright interference colors in thin section). When heated in a test tube it characteristically decomposes into numerous small white scales. It will not dissolve in acid (including sulfuric) or in KOH; it does dissolve in hot NaOH. In sulfuric acid it decomposes with difficulty, and only after strong heating. Diaspore is present in a number of bauxites.

Boehmite — AlOOH. This mineral has the same chemical composition as diaspore; the crystal system is orthorhombic; in crystal structure it is similar to lepidocrocite, FeOOH. The hardness is 3.5; the specific gravity ranges from 3.01 to 3.02-3.06. The mineral is white with yellowish tints, or it is colorless. The indices of refraction are Ng = 1.651-1.661, Np = 1.638-1.646; Ng−Np = 0.013-0.015; the average refractive index for cryptocrystalline varieties of boehmite is 1.640-1.645; the optic sign is positive.

Boehmite is distinguished from diaspore by lower hardness, by the optical properties (noticeably smaller birefringence and lower indices of refraction), and by the structural parameters; the differential thermal curves of boehmite and diaspore, on the other hand, are practically identical. Boehmite and diaspore are easily distinguished by immersion studies and by x-ray powder examinations. Boehmite does not dissolve in acid.

With hydrargillite boehmite is an essential constituent of many bauxites in the USSR. It makes up the principal bauxite deposits of France.

Hydrargillite (gibbsite) — $Al(OH)_3$ or $Al_2O_3 \cdot 3H_2O$. This is a so-called trihydrate of alumina, with the chemical composition of 65.4% Al_2O_3 and 34.6% H_2O; the system is monoclinic. The hardness is 2.5-3.5, the specific gravity 2.3-2.4. The color is white or grayish, light greenish, or reddish white. The indices of refraction are Ng, from 1.558-1.576 to 1.595 (generally 1.587); Np = Nm, from 1.535-1.554 to 1.581 (generally 1.566-1.568); Ng−Np = 0.014-0.030 (generally 0.019-0.021); the mineral is biaxial and positive; $c \wedge Ng = 21°$.

Hydrargillite is characterized by perfect cleavage in one direction and by a lower specific gravity (in comparison with other crystalline aluminum hydroxides); it is distinguished from diaspore by lower hardness, from kaolinite by greater birefringence. It dissolves in surfuric and hydrochloric acids and in hot NaOH. On heating it begins to change to boehmite, AlOOH (300-360°); at about 500-570° further dehydration occurs, and on heating to 950-1000° it changes into γ-Al_2O_3 (isometric system). The first two effects (of which the first is the greatest) are endothermic; the third is exothermic.

Hydrargillite is present in many bauxites in the USSR, both in crystalline form and in colloidal alumogel (gibbsite alumogel). Hydrargillite is widespread in normal sedimentary (subaqueous) sedimentary bauxite deposits (for example, in the Mesozoic bauxites on the eastern slope of the Urals) and in the clay-alumina products of lateritic weathering (subaerial accumulations); it sometimes forms from the activity of sulfuric-acid solutions.

Magnesium Oxides and Hydroxides

Brucite — $Mg(OH)_2$. This is magnesium hydrate, crystallizing in the trigonal system. It sometimes contains isomorphous admixtures of iron or manganese. The hardness is 2.5, the specific gravity 2 30-2 45. The crystals have a hexagonal-tabular aspect, frequently occurring in radiating lamellar aggregates. The color is white, rarely greenish or greenish yellow; the mineral is sometimes colorless; cleavage is perfect along (0001). The indices of refraction are Ne = 1.580-1.582, No = 1.559-1.562; Ne−No = 0.020-0.021; the mineral is uniaxial, positive. It occurs in tabular and platy crystals, in lamellar or radiating and fibrous aggregates. It is known to occur in metacolloidal incrustations. Normal-fibrous varieties are called nemalite; fibrous varieties may be biaxial.

Brucite is distinguished from alunite by better cleavage, from gypsum and muscovite by unixial figures and by positive sign, from hydromagnesite by generally greater birefringence. Brucite quickly dissolves in acid without effervescence; in particular it dissolves easily in hydrochloric acid, a fact that distinguishes it from hydrargillite.

Brucite forms in a strongly alkaline environment, by hydrolysis of soluble magnesium compounds, specifically: in soda lakes and highly alkaline soils, in weathering zones on serpentinites (especially in serpentinite masses), as a hydrothermal apomagmatic mineral that forms with serpentine during the hydrolysis of olivine; brucite is frequently found with hydromagnesite. It is also known in dolomitic limestones (with calcite, hydromagnesite, and periclase—MgO), supposedly as a hydrothermal mineral.

Iron Oxides and Hydroxides

Minerals of this group generally found in sedimentary rocks are magnetite, hematite, hydrohematite, goethite, hydrogoethite, turite, and lepidocrocite.

Magnetite (magnetic iron ore) — $FeFe_2O_4$ or $Fe_2O_3 \cdot FeO$. This is a strongly magnetic iron-ore mineral, with a chemical composition of 31% FeO and 69% Fe_2O_3 (72.4% Fe). It crystallizes in the isometric system. The hardness is 5.5-6.0, the specific gravity 5.0-5.2. The crystal form is octahedral or dodecahedral. The color is iron black, sometimes with a dark bluish iridescence. The mineral is opaque, bluish black in reflected light in thin section; under the same conditions in polished section, the mineral is dark blue or grayish blue, with a rose-brown cast; it is isotropic and is strongly etched with hydrochloric acid. Magnetite is generally easily distinguished by its strong magnetic property, by its black streak, by its color in reflected light, and by its crystal form. It is most easily distinguished from ilmenite by being more magnetic. Powdered magnetite dissolves in hydrochloric acid. The mineral is found in magmatic rocks (including dike rocks and ore deposits), in sedimentary iron ores, and as a mineral in regional and contact metamorphism. In sedimentary iron ores, magnetite is found in oolitic bodies— pisolites and oolites — and in epigenetic crystalline segregations along fractures and in cavities.

Magnetite of obviously sedimentary origin has been described by the author in iron ores of the Khalilova type, both in pisolites and in the special type of these ores — martite-chrome spinel — chlorite ores [Teodorovich, 1939₁]. It was noted by B. P. Krotov [1940] in cavities and fractures in this type of ore as having precipitated during epigenesis from migrating vadose waters; the same author also discovered the mineral in the same ores having formed pseudomorphs after wood fragments.

Pseudomorphs of hematite after magnetite commonly form during surface weathering in a hot climate; i.e., so-called martite is developed.

Hematite (red iron ore) — Fe_2O_3. The chemical composition is 70% Fe and 30% O. Hematite, or, more properly α-Fe_2O_3, crystallizes in the trigonal system (in contrast to the unstable polymorphous magnetic modification γ-FeO_3, or maghemite, which is in the isometric system). Hematite that is clearly crystalline and has a metallic luster is called iron-glance (specular hematite), whereas pseudomorphous hematite after magnetite is called martite.

The hardness is 5.5-6.0, the specific gravity 4.9-5.3. Crystals are tabular, platy, and rhombohedral; the mineral also forms dense masses. In crystalline varieties the color is iron black to steel-red; in earthy varieties (hydrohematite ?) the color is bright red. The streak is cherry red or red. Hematite is nonmagnetic; martite sometimes exhibits magnetism if remnants of magnetite have been preserved in it.

In polished sections hematite is grayish white with light blue tones (steel gray) or grayish white and white; the hardness is great; the mineral is slightly anisotropic; internal reflection is dark red. In thin sections under natural reflected light, the mineral is bluish gray or dark bluish gray with metallic luster; in reflection from electric lights, the mineral is gray with bluish tones and with metallic luster. Hematite forms dense masses of nodules that are cryptocrystalline or very fine grained.

Hematite is distinguished from similar minerals by streak, commonly by the form of the grains (tabular, flaky), by the color in thin sections and in polished sections in reflected light, by very slow solution in hydrochloric acid (concentrated), and by the absence of magnetism.

Hematite is widespread in ancient ferruginous quartzites, in nonmetamorphosed or slightly metamorphosed sedimentary iron ores (particularly in oolites and pisolites), and as a product of oxidation, especially during weathering in hot, arid climates.

Hydrohematite — α-$Fe_2O_3 \cdot aq$ or $Fe_2O_3 \cdot (0.01-0.9) H_2O$. A low-water iron hydroxide. It is a solid solution of water in hematite. The water content is less than 10%. The streak of the mineral is cherry red or cherry brown. The mineral is opaque in transmitted light in thin sections, but it may be translucent along the margins in very thin sections, showing a blood red color; in reflections of natural light, the mineral is dark bluish gray with metallic luster; in reflections of electric light, the color is a deep gray with dark bluish tones, and the luster is metallic. Polished sections show a grayish white or cream-white color in reflected light; they also exhibit well-defined anisotropism and pleochroism. The mineral is acid-resistant (dissolving slowly only in concentrated hydrochloric acid). It is present in a number of sedimentary iron ores.

Goethite — $HFeO_2$ or $Fe_2O_3 \cdot H_2O$. Iron oxide with a single molecule of water. Pure goethite contains 89.9% Fe_2O_3 and 10.1% H_2O, but the water content is generally higher than the theoretical value, corresponding to hydrogoethite. The mineral crystallizes in the orthorhombic system. In the goethite structure, all the oxygen ions in the lattice are chemically equivalent (in contrast to lepidocrocite). The hardness of goethite is 5.0-5.5, the specific gravity 4.00-4.28-4.50. The color is yellowish red or blackish brown; the streak is brown with reddish tones. The color in thin section is orangish or orange. The indices of refraction are Ng = 2.400-2.415, Nm = 2.39-2.409, Np = 2.26-2.275; Ng−Np = 0.14; the mineral is biaxial and negative and is slightly pleochroic, by which it may be distinguished from lepidocrocite. The specific gravity and the indices of refraction decrease in hydrogoethite with an increase in water content. In reflected light in polished sections, the color is light gray or gray, and anisotropism is distinct, with reddish brown internal reflection. The mineral dissolves slowly in hydrochloric acid.

Hydrogoethite — $HFeO_2 \cdot aq$ or $Fe_2O_3 \cdot (1\text{-}1.5)H_2O$. This mineral is an iron hydroxide, consisting of a solid solution of water in goethite. The water content in hydrogoethite ranges from 10 to 14.5%; the specific gravity is generally 4.0-3.6. The color is brown of various tones to black-brown. In transmitted light in thin section, hydrogoethite appears as an opaque dense mass with sections of finely dispersed structure or as translucent orangish fibers; this latter variety affects polarized light. The indices of refraction of distinctly crystalline varieties of hydrogoethite (fibrous and other forms) are Ng ≅ 2.36, Nm ≅ 2.35, and Np ≅ 2.21; pleochroism is from light yellow along Np to brownish yellow along Nm and orange-yellow along Ng. Interference colors are generally greenish. In thin sections in reflected light, the color of hydrogoethite ranges from orangish or brownish to dark brownish gray. In polished sections the color is light gray, sometimes with purplish or light bluish tones.

Hydrogoethite is distinguished from goethite and other similar iron minerals by specific gravity, indices of refraction, and water content; it may be distinguished from other hydroxides by its streak, by its common colloidal occurrence, by its color in reflected light, and, frequently, by being transparent in thin sections. It dissolves more quickly than goethite in hydrochloric acid. Hydrogoethite is widespread in sedimentary iron-ore deposits and in the oxidation zone of various deposits.

All iron hydroxides of the composition $HFeO_2 \cdot (0.01\text{-}1.5)H_2O$, or $Fe_2O_3 \cdot (1\text{-}4)H_2O$, are commonly called hydrogoethite: hydrogoethite proper (brown botryoidal ore), $HFeO_2 \cdot \frac{1}{6}H_2O$ or $Fe_2O_3 \cdot \frac{4}{3}H_2O$; limonite, $HFeO_2 \cdot \frac{1}{4}H_2O$ or $Fe_2O_3 \cdot \frac{3}{2}H_2O$; xanthosiderite, $HFeO_2 \cdot 0.5H_2O$ or $Fe_2O_3 \cdot 2H_2O$; limnite $HFeO_2 \cdot H_2O$ or $Fe_2O_3 \cdot 3H_2O$; and esmeraldaite, $HFeO_2 \cdot 1.5H_2O$ or $Fe_2O_3 \cdot 4H_2O$. There is doubt concerning the existence of some of these minerals. Isotropic colloidal iron hydroxides corresponding in composition to hydrogoethite is commonly called ehrenwerthite.

Iron ore ranging from brown-yellow to black and consisting of hydrous iron oxides of the composition $Fe_2O_3 \cdot (1.0\text{-}1.5)H_2O$ is ordinarily called brown iron ore or limonite. All the cryptocrystalline hydroxides of the hydrogoethite group (in the widest sense) and the lepidocrocite group are also called limonite.

Turite.* This term was previously used for a brownish red or red-brown iron mineral with the probable average composition of $2Fe_2O_3 \cdot H_2O$. The specific gravity is 4.1-4.6. It was later discovered that turite is a fine intimate intermixture (fine-grained or very fine-grained physical mixture) of two iron minerals — hydrohematite (sometimes hematite also) and goethite-hydrogoethite. This discovery was demonstrated by differential thermal curves and by x-ray studies; the relationship is also clearly seen in reflected light under the microscope. So-called red botryoidal ore consists of an intergrowth of parallel fibers, chiefly of hydrohematite and partly of goethite-hydrogoethite. Turite frequently is found in iron ores, in siderite and other ores, for example, which are products of secondary oxidation.

---

* Translator's note: Turgite is a more common form than turite, but both are considered more or less synonymous with hydrohematite. Since the author of this book has listed hydrohematite separately (above), no further effort has been made to resolve the usage of these terms.

Lepidocrocite (rubinglimmer) — FeOOH. This mineral is orthorhombic. During dehydration of lepidocrocite γ-Fe_2O_3 is formed, that is, the unstable and magnetic modification of Fe_2O_3, maghemite; during dehydration of α-Fe_2O_3 is formed, that is, the nonmagnetic variety of Fe_2O_3, hematite. Two types of oxygen ions are found in the lattice structure of lepidocrocite, and the formula of this mineral has come to be written FeOOH, although hydroxyl as such is not found in its structure. The hardness of lepidocrocite is 4, the specific gravity 4.09-4.10. The mineral occurs in platy crystals, but is found more frequently in fine-flaky and fibrous aggregates, which commonly show a parallel arrangement of the flakes and fibers. The color of the mineral is dark red to reddish brown; the streak is orange-red or brick-red. The mineral is transparent in thin sections, the color reddish orange or orange-yellow. The indices of refraction are Ng = 2.51, Nm = 2.20, and Np = 1.94; Ng−Np = 0.57. Pleochroism is marked: yellow or light yellow along Np, red-orange or orange along Nm, and red or orangish red along Ng.

Lepidocrocite is distinguished from goethite and hydrogoethite by strong pleochroism, from other iron oxides by its streak, from hematite by lower specific gravity, and from goethite by dissolving more rapidly in hydrochloric acid. In fine-crystalline and colloidal varieties, lepidocrocite may be distinguished from goethite only by x-ray methods.

Normally lepidocrocite forms concentric layers of scaly aggregates in sintery brown iron ores or on the walls of geodes, being here of surficial origin; in such occurrences one frequently finds alternating layers of lepidocrocite and parallel-fibrous goethite. In addition, lepidocrocite is known to be one of the latest of hydrothermal minerals. As there are goethite and hydrogoethite, so also is there (corresponding to lepidocrocite) hydrolepidocrocite, FeOOH · aq, which contains adsorbed water (a solid solution of water in lepidocrocite) and which is fine-crystalline or colloform masses.

According to A. G. Betekhtin [1950] and other authors, lepidocrocite and hydrolepidocrocite are much more widespread in nature than is generally believed.

Manganese Oxides and Hydroxides

In this subgroup we shall consider the following minerals that are frequently of sedimentary (including residual) origin: braunite, hausmannite, pyrolusite, manganite, vernadite, psilomelane and wad, and rancieite.

Braunite. This is manganese oxide with the approximate composition $Mn^{··}Mn_2^{···}O_3$, or $MnO · MnO_2$, frequently indicated by the abbreviated formula Mn_2O_3; the mineral generally contains silica (up to 8%). A. N. Winchell and H. Winchell [1953] adopted $(Mn,Si)_2O_3$ as the formula for braunite. According to A. G. Betekhtin [1950], silica occurs in braunite either as a finely disseminated mechanical admixture or by entering into the composition of the mineral; in the latter case the formula for braunite becomes $Mn(Mn, Si)O_3$. The chemical composition of pure braunite is 44.8% MnO and 55.2% MnO_2. The mineral crystallizes in the tetragonal system. The hardness is 6, the specific gravity 4.7-5.0; the mineral is nonmagnetic. The color is black, the streak brownish black. The luster is submetallic. The mineral is opaque. In polished sections in reflected light, braunite is grayish white and weakly anisotropic; rarely one may see cinnamon-brown internal reflection.

Braunite is distinguished from other minerals in the manganese oxide group by its great hardness and its black streak; it may be reliably identified in polished sections or by x-ray studies. It decomposes with difficulty in hydrochloric acid, with the separation of chlorine and gelatinous silica. In nitric acid it breaks down into MnO and MnO_2.

Braunite is known to occur in considerable quantities in sedimentary manganese deposits that have been subjected to regional metamorphism; it is also present in contact-pneumatolitic deposits and in hydrothermal veins.

Hausmannite — $MnMn_2O_4$ or $2MnO · MnO_2$ (abbreviated to Mn_3O_4). The chemical composition of hausmannite is 62% MnO and 38% MnO_2 (72% Mn). The mineral is tetragonal. The hardness is 5, the specific gravity 4.7-4.9; the mineral is nonmagnetic. The color is black, the streak cinnamon brown or red-brown. The luster in unoxidized varieties is adamantine or submetallic. Hausmannite is semitransparent; in transmitted light it is dark red-brown (No = 2.46, Ne = 2.15, uniaxial, negative), but oxidized or scarcely oxidized varieties are opaque. In reflected light in polished sections, the mineral is grayish white, strongly anisotropic, with red internal reflection.

In fine-grained masses, hausmannite is distinguished from other manganese oxides generally by microscopic examination or x-ray study. It is distinguished from braunite and hematite by lesser hardness, and from braunite also by the streak. Microscopically, hausmannite is recognized by distinct optical anisotropism, by red internal reflection, and by polysynthetic twinning, as seen on grains in polished sections. The mineral dissolves in hydrochloric acid, with the separation of chlorine. In nitric acid it decomposes slowly.

Hausmannite occurs in large masses with braunite in regionally metamorphosed sedimentary manganese deposits; it is also found in some contact-metasomatic and hydrothermal manganese deposits. It forms in a more distinctly reducing environment than braunite, in places forming pseudomorphs after the latter.

Pyrolusite — MnO_2. This mineral contains 63.2% Mn; the crystal system is tetragonal.

Polianite is a distinctly crystalline variety of pyrolusite, but some authors consider the two terms to be generally synonymous. Pyrolusite normally occurs in compact crystalline or cryptocrystalline masses, frequently in earthy powdery masses, to some extent in concretions, and rarely in crystals (acicular or prismatic). The hardness of crystals is 5-6, but in aggregates this value is as low as 2; the specific gravity is 4.7-5.0. The color and streak are both black. Luster is submetallic. Cleavage is perfect along (110). The mineral is opaque. In reflected light in polished sections, the color is cream or gray-white with cream tints, or it is cream-white; anisotropism is marked.

Pyrolusite is distinguished from other manganese oxides by black streak, characteristic cleavage, brittleness, and, in aggregates, low hardness. On heating at a temperature of 550-650° the mineral changes to braunite, and at 940-1100° to hausmannite. It dissolves in hydrochloric acid, with the separation of chlorine.

Pyrolusite forms in a strongly or distinctly oxidizing environment, in littoral marine zones and lake basins, or in the oxidation zones of manganese deposits.

It is found in sedimentary manganese deposits, especially of Tertiary age (Chiatura in Georgia, Nikopol' in the Ukraine, Polunochnoe in the Urals, and others), where, with psilomelane, it forms oolites (Fig. 27),

Fig. 27. Oolites of psilomelane and pyrolusite, sooty pyrolusite in the cement between them; polished section, × 18. (From Betekhtin, 1937.)

concentrically zoned concretions, and local distinctly crystalline masses (the Akkermanovo deposit in the Urals). Pyrolusite also occurs in all oxidation zones of manganese deposits (hydrothermal and other types). It commonly forms pseudomorphs after manganite, vernadite, psilomelane, hausmannite, and other minerals.

Manganite — $Mn^{..}Mn^{...}O_2(OH)_2$ or $MnO_2 \cdot Mn(OH)_2$. The chemical composition is 40.4% MnO; 49.4% MnO_2, and 10.2% H_2O. Of the impurities that may be present, silica should be mentioned (it may constitute up to several percent of the mineral mass). The water content in cryptocrystalline and oolitic varieties may be greater than that indicated by the formula (hydromanganite); this excess water is adsorbed water. Manganite is monoclinic, crystals having a prismatic aspect. The hardness is 3-4, the specific gravity 4.2-4.35 (in hydromanganite this value may be as low as 3.7). The color is black, though varieties rich in water are brown; the streak is black. The mineral is opaque, translucent in very thin sections. The indices of refraction are Ng = 2.53, Nm = 2.24, Np = 2.24; Ng—Np = 0.29. In reflected light in polished sections, the color is grayish white to gray; anisotropism is very marked, with blood-red or brownish red internal reflection. Manganite dissolves in hydrochloric acid, with the separation of chlorine.

Manganite and hydromanganite are widespread in sedimentary manganese ore deposits of Tertiary age, occurring in oolites (Chiatura and other deposits) and in compact masses. The minerals generally occur in the intermediate zone between carbonate manganous-oxide ores and psilomelane-pyrolusite manganic-oxide ores. In clays manganite is sometimes found in nodules with radiating structure. It is also known to occur as a late hydrothermal mineral.

Vernadite — $MnO_2 \cdot nH_2O$. A hydrous manganese oxide, occurring in weakly crystallized or in colloidal masses; it is rich in H_2O. The MnO_2 content is 70-82%; water generally constitutes 12 to 8%; CaO is present (sometimes up to several percent) as an impurity, and MnO generally makes up 1-2%. The hardness is 2-3, the specific gravity 2.28-3.00. Dense varieties are pitch-black, but powdery varieties are chocolate-brown; the streak is chocolate-brown or reddish brown. Reliable identification may be made only by means of chemical analysis. Almost all the water is gradually eliminated from vernadite on heating up to 120°. The mineral is easily dissolved in hydrochloric acid, with the separation of chlorine.

According to A. G. Betekhtin, vernadite forms during hydrolysis and oxidation chiefly of carbonates (rhodochrosite and others) and silicates (rhodonite — $MnSiO_3$ — and others) of bivalent manganese. Vernadite, because of the color of its streak, was previously often mistaken for manganite. During dehydration vernadite loses water (varieties are known containing from 8 to 5%) and changes to pyrolusite.

Psilomelane — $mMnO \cdot MnO_2 \cdot nH_2O$. The indicated formula is approximate. In the American literature a variety with essential barium (romanechite of other authors), with the formula $BaMn^{..}Mn_8^{...}O_{16}(OH)_4$, is called psilomelane. The ratio of MnO_2 to MnO is variable. The mineral generally contains 60-80% MnO_2, 8-25% MnO, and 4-6% H_2O. BaO is commonly present in small quantities (up to several percent). The crystal system is orthorhombic (?). The hardness is 4-6, the specific gravity 4.4-4.7. Psilomelane generally forms incrustations (concentrically zoned), fine-crystalline varieties, and also oolites and concretions; it is sometimes found in typical colloidal (x-ray amorphous) varieties — wad. The color is black, sometimes brownish black; the streak is black. The luster is submetallic (friable varieties are dull). In reflected light in polished sections, psilomelane is grayish white to light white-gray; it is isotropic, sometimes showing cinnamon-brown internal reflection.

Psilomelane is characterized by sintery appearance, fine-crystalline and oolitic occurrences, black streak, and by the fact it belongs to the first group of manganese oxides according to the reaction of K. Fadeev (see p. 79). A reliable identification requires chemical and x-ray analyses. The mineral dissolves in hydrochloric acid, with the separation of chlorine.

Psilomelane forms in the surface zones of the sedimentary shell (in an exogenetic environment), in deposits of sedimentary origin (Chiatura, Nikopol', and other deposits) and in the oxidation zones of manganese ore deposits; it is known as a secondary mineral in manganese ores of hydrothermal origin. In sedimentary manganese deposits, psilomelane is found in seams of dense ore or as oolites (see Fig. 27) and spherical concretions. During weathering psilomelane is oxidized and dehydrated, changing to pyrolusite.

Wad — $mMnO \cdot MnO_2 \cdot nH_2O$. This is generally an earthy or sintery-colloidal mineral (earthy amorphous psilomelane). Some authors consider wad to be a mixture, consisting chiefly of pyrolusite and psilomelane. The hardness is 1-3, the specific gravity 2.3-3.7. The color is brownish black. Wad is a manganese ore of exogenetic origin. It is known locally in Tertiary deposits in the southern part of the European USSR and in other regions.

Rancieite — $m(Mn, Ca)O \cdot MnO_2 \cdot nH_2O$. This is calcium psilomelane, in which the content of CaO is as much as 9% and H_2O constitutes 12-13%. The crystal system is unknown, but the x-ray powder photographs of rancieite are different from the pictures of the other manganese oxides. This mineral is found in compact masses (fine-platy or fibrous); in addition, it occurs in incrustations and in nodular forms. The color is black or steel-gray, but in a finely dispersed form it is cinnamon brown; the streak is cinnamon brown. The luster is metallic. The mineral is opaque. The hardness is 2-3, the specific gravity 3.3-3.4. In polished sections rancieite appears anisotropic.

This mineral is characteristic in the zones of oxidation or semioxidation in manganese ore deposits, particularly in sedimentary deposits (Chiatura, Polunochnoe, and others).

* * *

Minerals in the group of manganese oxides and hydroxides may be separated by means of the reaction of K. Fadeev [1892] into two subgroups: those that stain a solution and those that do not.

The Fadeev reaction consists of the following: 1) a mixture of one part distilled water and one part concentrated sulfuric acid are placed in a test tube (be sure that H_2SO_4 is poured into the water, and not vice versa), and into this is introduced a small portion of the powdered manganese mineral; 2) the contents of the tube are brought to boiling and boiled (being held in a stand) for several seconds or a minute*; the test tube is then removed from the burner. Two results are possible: a) the solution (after the suspension has settled, or with material still in suspension if there is a little of it) becomes a rose-purple (generally weak) color; braunite, hausmannite, psilomelane, and wad react in this way; b) the solution is not stained (after the suspension has settled); the minerals in this subgroup include pyrolusite, manganite, polianite (variety of pyrolusite), and vernadite. The indicated reaction facilitates the identification of minerals in the group of manganese oxides and hydroxides, constituting an auxiliary criterion.

Copper Oxides

Among the sedimentary authigenic minerals in this subgroup are cuprite and tenorite.

Cuprite (red copper ore) — Cu_2O. This is a copper oxide with the chemical composition of 88.8% Cu and 11.2% O, crystallizing in the isometric system. The hardness is 3.5-4.0, the specific gravity 5.85-6.15. Crystals are octahedral, rarely some other form. The color is red to lead-gray (in fine-grained and cryptocrystalline aggregates). The streak is cinnamon- or brownish-red. The mineral is semitransparent in thin splinters or in petrographic thin sections; it has an index (N) of 2.85 and is isotropic. In reflected light in polished sections, cuprite is grayish white, clearly anisotropic, and shows ruby-red internal reflection.

Cuprite is characterized by its red streak and its association with native copper, malachite, and azurite. Under the microscope it is distinguished from other red minerals (when it is semitransparent) by its isotropic character. Curpite dissolves easily in nitric acid; in the process the solution becomes green, but is changed to blue on adding an excess of ammonia.

Cuprite is generally found in compact granular, sometimes earthy, dirty aggregates, rarely in small crystals or inseparable acicular aggregates. It occurs with native copper, malachite, azurite, and other secondary copper minerals. It generally forms during exogenetic processes — oxidation of chalcocite and other sulfide ores of copper. When the CO_2 content becomes large in solutions, cuprite becomes unstable and is generally replaced by malachite, rarely by azurite.

Cuprite, having formed by oxidation of chalcocite, is found in two forms in the typical sedimentary cupriferous sandstones in the Permian rocks of Tataria [Miropol'skii, 1938]: as monomineralic incrustations and efflorescences and as earthy concretionary nodules in intimate association with hydrogoethite and with included clastic impurities.

Tenorite — CuO. This is copper oxide that crystallizes in the monoclinic system. The hardness is 3.5, the specific gravity 5.8-6.4. This mineral forms thin scales, bar-like prisms, or earthy masses. The color is black or grayish black, the streak grayish black. In transmitted light tenorite is transparent, showing pleochroism (from dark brown along Ng to light brown along Nm). In reflected light in polished sections, the color is grayish white to yellow-brown; anisotropism is weak. The mineral dissolves easily in acid.

_____

*If the heating is continued and the contents let stand afterward, all the manganese minerals in the investigated group will stain the solution.

Tenorite is found in oxidized zones of copper sulfide deposits, together with cuprite, hydrogoethite, malachite, chrysocolla, and other minerals; it also occurs in typical sedimentary cupriferous sandstones, such as Permian sandstones [Miropol'skii, 1938], as an earthy variety.

5. SULFIDE GROUP

Iron and copper sulfides are widely found in sedimentary formations; such formations also contain, in very small quantities. sulfides of manganese, lead, and zinc. It is quite obvious that sulfides can form only in a markedly alkaline or distinctly alkaline environment, that is, where oxygen is absent and hydrogen sulfide is present.

Iron Sulfides

Normal sedimentary iron sulfides are hydrotroilite, melnikovite, pyrite, marcasite, and pyrrhotite; the sulfide of iron and copper, chalcopyrite, is also present.

Hydrotroilite. This is a hydrosulfide of iron, being a black finely dispersed colloidal substance, and is found in Recent marine muds and in the sediments and muds of estuaries and lakes. Hydrotroilite is apparently a hydrogel of iron monosulfide (i.e., troilite, FeS) with adsorbed water — $FeS \cdot nH_2O$ [M. Sidorenko, 1907, 1909]. A. D. Arkhangel'skii [1934] pointed out the wide development of hydrotroilite in Recent sediments in the Black Sea, especially at shallow and moderate depths; he also indicated its ability to migrate into layers underlying Recent sediments.

Hydrotroilite is also known to occur in older deposits — limestones and clays.

In the course of time hydrotroilite (through melnikovite) changes to pyrite or marcasite, rarely to pyrrhotite.

Melnikovite — FeS_2. This is a colloidal or metacolloidal cryptocrystalline finely dispersed variety of iron disulfide, black or gray-black in color. X-ray powder photographs in some cases show a pattern of pyrite; others show a marcasite pattern. This fact leads one to believe that melnikovite is a transitional product between hydrotroilite and pyrite or marcasite.

The mineral is known in sedimentary rocks (clays, rarely in marls, sandstones, and other rocks), ores (carbonate manganese ores), and muds; it is precipitated from hot springs, and it occurs in ore deposits (i.e., it may be sedimentary or hydrothermal).

Pyrite (iron pyrites) — FeS_2. An iron disulfide with the composition 46.6% Fe and 53.4% S; it is isometric (in contrast to the orthorhombic marcasite). The hardness is 6.0-6.5, the specific gravity 4.9-5.2. Crystals are most commonly cubical or pentagonal dodecahedral (pyritohedral). The color is light brass-yellow, commonly showing a yellowish brown or variegated tarnish; the streak is brownish or greenish black. The luster is strongly metallic. In reflected light in thin sections, pyrite is brass-yellow with metallic luster; in polished sections the color is creamy white or yellowish white; the mineral is isotropic (marcasite, which is similar to pyrite, is lighter and anisotropic), and it is etched with nitric acid.

Pyrite is found in sedimentary rocks in a variety of forms: disseminated cubes and pyritohedrons, rarely octahedrons and diploids; rounded and irregularly formed microaggregates; occasionally as cement in sandstones, concretions, etc. Pyrite is distinguished from marcasite, chalcopyrite, and pyrrhotite by greater hardness (it scratches glass), by striated faces parallel to the rhombs (100) and (210), by crystal form, and by color. In reflected light pyrite is somewhat darker than marcasite, and in polished sections it is distinguished by its isotropic character. It will not dissolve in weak hydrochloric acid, but it dissolves with difficulty in nitric acid (on heating). When it is boiled in a 3% solution of $AgNO_3$, it merely becomes slightly brown, whereas marcasite is stained a darker color, later changing to red and, finally to blue.

Pyrite is the most widespread sulfide in the earth's crust; it is abundant in sedimentary rocks, is found in numerous magmatic rocks, in contact-metasomatic and hydrothermal deposits, and is also present in slates; it frequently occurs as pseudomorphs after marcasite or melnikovite. It forms in a hydrogen sulfide environment. The pyrite (or marcasite) period of diagenesis of sediments is of great importance in distinguishing sedimentary geochemical and mineral-geochemical facies: when benthonic forms are missing it identifies the sulfide or hydrogen sulfide facies, and with siderite (or chamosite) it characterizes the sulfide-siderite (sulfide-chamosite) facies. As is well known, most adherents of the organic theory of petroleum genesis consider the sulfide and sulfide-siderite marine facies, and saline or brackish waters in general, as the probable petroleum-producing environments, where the deposits contain initially considerable quantities of organic sapropelic material.

During surface or deep weathering, pyrite is oxidized, changing to iron hydroxides and sulfuric acid; when $Ca(HCO_3)_2$ is present in solution, gypsum is formed.

Marcasite — FeS_2. This is iron disulfide, crystallizing in the orthorhombic system (in contrast to pyrite). The hardness is 5-6, the specific gravity 4.6-4.9. Crystals are tabular or short-prismatic with spear-shaped terminations. The orthorhombic character of marcasite may sometimes be noted in aggregates of radially arranged spear-shaped crystals and is frequently observed in the radiating structure in concretions and microconcretions; but marcasite is most frequently found in sedimentary rocks in rounded and irregularly formed microaggregates. The color is pale bronze-yellow to grayish-yellow or brass-yellow, with greenish or grayish tones. The streak is dark greenish gray. Marcasite is opaque and has a marked metallic luster. In reflected light in thin sections, the color is brass-yellow; in polished sections it is creamy white.

Marcasite is distinguished from pyrite, apart from the crystal form, by lesser hardness and lower specific gravity. It should be noted that pseudomorphs of pyrite after marcasite have been found, some of them after crystals of distinctively orthorhombic habit. It is therefore advisable here to introduce, in addition to the indicated criteria, other features by which marcasite may be distinguished from pyrite: 1) in reflected light marcasite is lighter than pyrite; 2) in polished sections pyrite is isotropic,* but marcasite shows anisotropic colors from gray or greenish gray to rose-gray; 3) in concentrated hydrogen peroxide (H_2O_2), pyrite yields a white flocculent precipitate of colloidal sulfur, but marcasite gives a clean solution or a whitish clayey suspension; 4) the x-ray powder photographs of pyrite and marcasite are different; 5) pyrite is etched only by nitric acid, but marcasite is affected by aqua regia as well; 6) when a small quantity of powdered pyrite is treated first with cold nitric acid and then, when violent reaction has ceased, heated, most of the sulfur is oxidized and goes into solution as SO_3, but when marcasite is treated in the same manner, most of the sulfur separates out in the free state; 7) when powdered mineral is boiled in 3% solution of $AgNO_3$, marcasite becomes tobacco brown, then red, and, finally blue, but pyrite only becomes slightly brown.

Marcastie is found in sedimentary deposits and also in hydrothermal (vein) deposits. Sedimentary marcasite is an authigenic mineral, being encountered chiefly in carbonaceous sandy clay deposits and in primary bituminous sequences of sand-silt-clay, clay, and marls; like pyrite, it indicates a distinctly reducing environment. Where marcasite formed during diagenesis, it may, like pyrite, identify either the sulfide or the sulfide-siderite geochemical facies in the absence of benthonic forms. Sedimentary diagenetic marcasite apparently forms by the crystallization of colloidal sulfides; it subsequently changes easily to pyrite. Finally, marcasite commonly develops from pyrrhotite in the lowest parts of the weathering zones of ore deposits; in such cases it is fine grained or, occasionally, colloform.

When marcasite is oxidized, as when pyrite is oxidized, iron hydroxides are formed, and sulfuric acid, which leads to the development of gypsum when $Ca(HCO_3)_2$ is present in solution.

Pyrrhotite (magnetic pyrites). Iron sulfide, ranging mostly from Fe_6S_7 to $Fe_{11}S_{12}$, and having the general formula $Fe_{1-x}S$, in which x ranges from 0 to 0.2; most frequent x = 0.1-0.2, but if x = 0, the mineral is called troilite and is generally designated by the formula FeS. The crystal system is hexagonal. The hardness is 4, the specific gravity 4.5-4.7. The color is dark bronze-yellow with a brown tarnish; the streak is grayish black. The luster is metallic. The mineral is magnetic, but the degree of magnetism varies. In reflected light pyrrhotite is rose-cream and is strongly anisotropic.

Pyrrhotite is found chiefly in igneous rocks, in contact-metasomatic and hydrothermal deposits, but sometimes with siderite in sedimentary formations (the Kerch iron-ore deposit), in phosphoritic nodules, and also in oolites (Mesozoic-Cenozoic rocks of Central Asia) and in Recent muds. In the oxidized zone, pyrrhotite is most easily decomposed of the sulfides.

---

* As has been discovered, most pyrite is weakly anisotropic (from light gray to gray), whereas a smaller proportion of pyrite is absolutely isotropic.

81

Copper Sulfides

Normal sedimentary copper sulfides are chalcocite, bornite, covellite, and the sulfide of copper and iron, chalcopyrite. We begin the descriptions with the most widespread — chalcopyrite.

Chalcopyrite (copper pyrites) — $CuFeS_2$. A sulfide of copper and iron with the chemical composition 34.57% Cu, 30.54% Fe, and 34.9% S; the crystal system is tetragonal. The hardness is 3.5-4.0, the specific gravity 4.1-4.3. The color is brass-yellow, commonly with a dark yellow or variegated tarnish. The streak is black with a greenish cast. The mineral is opaque. The luster is strongly metallic. In reflected light in polished sections, the mineral is yellow and weakly anisotropic. Chalcopyrite is found in compact masses, disseminations, and colloform deposits; it commonly contains pyrite as an impurity. It is distinguished from pyrite by lesser hardness, by lower specific gravity, and by a deeper yellow color; it is distinguished from marcasite by the first two features. Chalcopyrite dissolves in nitric acid, with the separation of sulfur.

Chalcopyrite is widely developed in hydrothermal vein and metasomatic deposits, is frequently found in magmatic copper-nickel (sulfide) deposits, and also occurs in zones of secondary sulfide enrichment in copper sulfide deposits and in sedimentary rocks. It is the most common and important copper mineral.

In sedimentary rocks chalcopyrite forms where hydrogen sulfide is present, from the decomposition of organic remains or from introduced solutions; it replaces wood and animal remains and other plant fragments, or it is precipitated without such material being present. In recent times chalcopyrite has been more and more frequently observed in normal sedimentary rocks, but in small quantities. It is known to occur in the cupriferous sandstones of the Donets Basin (Lower Permian), in late Paleozoic red cupriferous sandstones of Kazakhstan, very rarely in the Permian rocks of Tataria, and in very small quantities in various other deposits (the Devonian of Tataria and others). D. G. Sapozhnikov and I. P. Zlatogurskaya [1953] noted that copper mineralization occurred chiefly in Kazakhstan in gray sandstones, siltstones, and limestones.

Bornite (variegated copper ore) — Cu_5FeS_4. A sulfide chiefly of copper, crystallizing the isometric system. The hardness is 3, the specific gravity 4.9-5.0. The color on fresh fractures is dark copper-red, but bornite is generally covered with a variegated (chiefly dark blue) tarnish; the streak is grayish black. The mineral is opaque. In polished sections a characteristic brownish rose color may be recognized; the mineral is normally isotropic, but occasionally one may observe weak anomalous anisotropism.

Bornite is distinguished by its color, its blue tarnish, its low hardness, and its characteristic color in polished sections. It decomposes in nitric acid, with the separation of sulfur.

Endogenetic bornite (hydrothermal) and exogenetic bornite (in zones of secondary sulfide enrichment, chiefly at the site of chalcopyrite minerals) are known. Exogenetic bornite is frequently present in cupriferous sandstones, where it may be the principal copper mineral. The origin of bornite in cupriferous sandstones is variable: the mineral may be diagenetic, epigenetic, or it may form during metamorphism.

Bornite easily decomposes in air.

Chalcocite (copper glance) — Cu_2S. A copper sulfide with the composition 79.8% Cu and 20.2% S. This sulfide is the richest in copper. The crystal system is orthorhombic. The hardness is 2-3, the specific gravity 5.5-5.8. The color is bluish gray, the streak dark gray. The luster is metallic. In reflected light in polished sections, the mineral is grayish white or bluish white and slightly anisotropic. Chalcocite is found chiefly in compact fine-grained masses or as disseminations, pseudomorphs (after chalcopyrite, covellite, or other sulfides).

Chalcocite is distinguished by its bluish gray color, its low hardness, its malleability (sharp blades retain shiny marks), and by its good electrical conductivity. It dissolves in acid, especially well in nitric acid, with the separation of sulfur; solutions in nitric acid become green.

Chalcocite is chiefly an exogenetic mineral, forming in the zone of secondary sulfide enrichment of all copper sulfide deposits; it also forms in normal sedimentary subaqueous deposits and is occasionally a hydrothermal mineral.

Chalcocite precipitates in a hydrogen sulfide environment from copper-bearing solutions in lithified sedimentary rocks (by redeposition), or it forms in sediments (directly) as pseudomorphs after woody fragments or completely independent of such material.

Chalcocite is rather widespread in sedimentary rocks; L. M. Miropol'skii [1938] described it from Kazanian (in part from Tatarian and Artinskian*) deposits in Tataria; it has been found in cupriferous sandstones in the Donbas; V. A. Polyanin and I. N. Gorizontova [1939] discovered it in Permian (Kazanian) deposits in the Kirov oblast; D. G. Sapozhnikov and I. P. Zlatogurskaya [1953] found it in cupriferous Paleozoic sandstones in Kazakhstan.

L. M. Miropol'skii [1938] noted that, for the Kazanian series in Tataria, chalcocite occurs chiefly in sandy-marly clays, in which it forms concretions, nodules, clots, and small disseminated inclusions. It is found much less abundantly in sandstones in this region, but where it does occur in such rock it is associated with accumulation of plant debris.

Chalcocite is unstable in the oxidized zone; it changes to cuprite, malachite, or azurite.

Covellite (copper indigo) — CuS or $Cu_2S \cdot CuS_2$. A copper sulfide with the composition 66.5% Cu and 33.5% S; it is hexagonal. The hardness is 1.5-2.0, the specific gravity 4.59-4.67-4.76. The color is indigo-blue or dark blue-black, the streak gray to black. In ordinary thin sections covellite is opaque, but in very thin sections it is translucent and pleochroic in light green tones; it is uniaxial and positive. In reflected light the mineral is blue, strongly anisotropic, and pleochroic. It occurs in films, admixtures, and powdery and sooty masses.

Covellite is distinguished by its bright blue color, its low hardness, its occurrence with other copper sulfides, and its anisotropism and pleochroism in reflected light. It dissolves in hot nitric acid, the solution becoming green.

Covellite is a characteristic exogenetic mineral in the zone of secondary sulfide enrichment in copper-ore deposits, where it is generally found in very small quantities at the site of chalcopyrite, chalcocite, and other sulfides. It is unstable in the oxidized zone, changing to copper sulfate.

Covellite has been found in very small quantities in normal sedimentary rocks, for example, in cupriferous sandstones in the Donets Basin, in cupriferous Paleozoic sandstones of Kazakhstan, and, very rarely, in Permian deposits in Tataria and the Kirov oblast.

Manganese, Lead, and Zinc Sulfides

These sulfides are generally found in sedimentary rocks in very small quantities. They are alabandite, hauerite, galena, and sphalerite.

Alabandite (alabandine) — MnS. A manganese sulfide with the chemical composition 63.2% Mn and 36.8%; the crystal system is isometric. The hardness is 3.4-4.1, the specific gravity 3.9-4.1. The color is dark green and steel-gray to iron-black; the streak is greenish. Alabandite is found in irregular disseminations and imperfect crystals. In polished sections the color is gray; the mineral is isotropic, with dark green internal reflection. It dissolves quickly in hydrochloric and nitric acids, giving off an odor of hydrogen sulfide. Alabandite is known to occur in hydrothermal associations, in weakly metamorphosed sedimentary manganese carbonate ores, and occasionally in normal sedimentary rocks (the Lower Permian rocks of the Ishimbai region). It is unstable in air, oxidizing quickly.

Hauerite — MnS_2. A manganese sulfide with the composition of 46.2% Mn and 53.8% S; it crystallizes in the isometric system. The hardness is 4, the specific gravity 3.4-3.5. The color is brownish gray to brownish black, the streak brick-red or reddish brown. The mineral is opaque; in thin sections it is occasionally translucent, having a dark red color; it is isotropic in such occurrences. In polished sections hauerite is grayish white with slight brownish tones. It dissolves in hydrochloric acid, the solution taking on a cinnamon-brown color.

Hauerite forms in a distinctly reducing hydrogen sulfide environment during the diagenesis of manganese-bearing sediments; it develops in small quantities. It has been definitely recognized in the carbonate manganese ores of the Chiatura deposits and, in small quantities, in slightly metamorphosed sedimentary manganese ores in the eastern Transbaikal.

Galena (lead glance) — PbS. A sulfide of lead with the chemical composition 86.6% Pb and 13.4% S; the system is isometric. The hardness is 2-3, the specific gravity 7.4-7.6. The mineral is lead gray, the streak

---

*Sakmarian-Artinskian.

83

grayish black. The luster is metallic. Galena has perfect cubic cleavage. It is opaque. In reflected light in polished section, it is white, isotropic, with no internal reflection. In cryptocrystalline masses galena may be identified by its specific gravity and by its easy solubility in nitric acid (with the separation of sulfur and a white precipitate of $PbSO_4$); in general, the mineral is identified by its color, luster, cubic cleavage, low hardness, and specific gravity.

Galena is a characteristic mineral in hydrothermal deposits, where it frequently forms rich concentrations in association with sphalerite (ZnS). It is considered rare in sedimentary rocks, but recently it has been more and more frequently identified in such rocks, though in small quantities. Sedimentary galena is known in coal-bearing strata, in some phosphorites and spherosiderites (iron concretions) [Konstantinov, 1954] and in oil strata, where it may be diagenetic (most frequently) and epigenetic. Galena is an indicator of a clearly reducing environment.

Galena that has been found in sedimentary rocks is thought by some investigators to be hydrothermal in origin, even when no indications of hydrothermal activity can be found. Recent discoveries of exogenetic galena in the lower oxidized zones of sulfide deposits are therefore of considerable interest [Rudenko, 1954]. In these deposits galena forms coatings consisting of small grains about coarse primary grains of sphalerite; locally it fills fine fractures in sphalerite, commonly forming on fibrous smithsonite ($ZnCO_3$) veinlets that wedge out in the sphalerite. In this latter occurrence galena formed during partial solution of the sphalerite; this relationship raises the question concerning different values of the oxidation-reduction potential for the formation of PbS and ZnS, but the problem cannot be solved without consideration of the composition of the reacting solution itself.

When galena oxidizes in air or in water, it changes into sulfate ($PbSO_4$) and then to cerussite ($PbCO_3$); the coating thus formed, because of its low solubility, preserves the grains or parts of the galena from complete oxidation.

Sphalerite (zinc blend) — ZnS. A sulfide of zinc with the composition 67.1% Zn and 32.9% S; the crystal system is isometric. The hardness is 3-4, the specific gravity 3.9-4.2. Crystals have a tetrahedral aspect, sometimes dodecahedral. The color is brown-black, brown, black, rarely yellowish, sometimes greenish or reddish; the streak is white, light yellow, or light brown (ferruginous varieties have a cinnamon-brown streak). The luster is adamantine. Cleavage along (110) is perfect. The indices of refraction in translucent crystals (in petrographic thin sections sphalerite is transparent to translucent) is high: N = 2.37 increasing with increase in iron content to 2.43-2.47. In reflected light in polished sections, the mineral is gray; it is isotropic, with internal reflection from white to dark cinnamon-brown.

Sphalerite is distinguished by the equant form of the crystals (with cleavage along the dodecahedron) and by the luster. In thin sections it is recognized by marked positive relief, by its isotropic character (ordinarily, unless it is anomalous), and by its cinnamon-brown color; in immersions it is semitransparent (dark gray or cinnamon-brown grains). Sphalerite dissolves in concentrated nitric acid, with the separation of sulfur; it also dissolves in strong hot hydrochloric acid.

Sphalerite is chiefly a hydrothermal mineral. In sedimentary rocks it may be present as allogenic grains, but it has also been found in small quantities as an authigenic constituent (diagenetic and epigenetic). Authigenic sedimentary sphalerite is found in some coal deposits, some spherosiderites (iron concretions) and phosphorite concretions; it is rarely found in oil strata and in individual carbonate rocks.

6. PHOSPHATE GROUP

Sedimentary phosphates are rather widespread authigenic minerals; of these, two principal subgroups may be distinguished — calcium phosphates and iron phosphates.

Calcium Phosphates

Authigenic sedimentary calcium phosphates are chiefly associated with phosphorites, but are also found in sedimentary rocks in which phosphorites are not abundant. The following principal varieties of sedimentary calcium phosphates may be distinguished: fluorapatite, hydroxyapatite, fluorhydroxyapatite, francolite (staffelite), collophane, kurskite, and carbonate-apatite (podolite or dahllite).

One of two variants of a widely used qualitative reaction [Teodorovich, 1950, p. 49] is generally used to detect calcium phosphates. In the first, concentrated nitric acid and ammonium molybdate are the reagents used. First, powdered ammonium molybdate is washed in strong nitric acid to test its purity (during this washing the white color of the powder should not noticeably change). Then a small piece of the rock or mineral to be tested is placed in a beaker, sprinkled with ammonium molybdate powder, and dampened with strong nitric acid. The appearance of a yellow precipitate and the thickness of such precipitate indicate the presence and quantity of calcium phosphate, as well as the distribution of the material through the rock.

In the second variant, a large drop of ammonia is placed on a piece of the material to be tested, or better, on some of the powdered material; five drops of benzidine are added (which leads to a slightly yellowish coloration) and 2-3 drops of 10% solution of ammonium molybdate are introduced. Calcium phosphate gives a dark blue coloration.

Fluorapatite — $Ca_5(PO_4)_3F$ or $Ca_{10}P_6O_{24}F_2$. This is the more widespread of the two known varieties of apatite (fluor- and chlorapatite). The chemical composition is 55.5% CaO, 42.3% P_2O_5, and 3.8% F. The mineral crystallizes in the hexagonal system. The mineral $Ca_5(PO_4)_3(OH)$, also hexagonal, is called hydroxyapatite.

The hardness of fluorapatite is 5, the specific gravity 3.17-3.23. The color is pale green, green, light blue, yellow, purple, or white; it is sometimes colorless. The indices of pure fluorapatite are No = 1.633-1.634, Ne = 1.629-1.632; No−Ne = 0.005-0.002; the mineral is unixial and negative, extinction is parallel, and the elongation is negative. Fluorapatite dissolves in nitric, hydrochloric, and sulfuric acids.

In sedimentary rocks fluorapatite is found as an authigenic constituent as well as in fragmental, or allogenic, grains. Authigenic syngenetic fluorapatite forms some nodular or tabular phosphorites; these occur in marine sedimentary formations in which fluorapatite and hydroxyapatite are present. In particular, fluorapatite constitutes the principal concretionary Cambrian-Silurian phosphorites and the principal nodular Cenomanian phosphorites of Podolia [Furman, 1954]. Fluorapatite is also found in some igneous and metamorphic rocks, in pegmatites, and in hydrothermal veins.

Hydroxyapatite — $Ca_5(PO_4)_3(OH)$ or $Ca_{10}P_6O_{24}(OH)_2$. This is fluorapatite or chlorapatite in which the halogen has been replaced by hydroxyl; the crystal system is hexagonal. The specific gravity is 2.97-3.07. The color is white or wax-yellow. The indices of refraction are No = 1.651-1.657 and Ne = 1.644-1.653; No−Ne = 0.007-0.004. Fresh animal bones and teeth and, apparently, the excrement of marine birds (guano) are composed of hydroxyapatite. In addition, the mineral is present in talc-chlorite schists. In time the hydroxyapatite in bones and teeth changes to francolite.

The reworked Cenomanian phosphorites of Podolia are also composed of a hexagonal fluorhydroxyapatite mineral intermediate in composition between fluorapatite and hydroxyapatite (its indices are No = 1.645 and Ne = 1.641). The main bulk of the indicated phosphorites contain 2.04% F and 35.28 P_2O_5; in the zones of these reworked nodules there are small crystals of almost pure hydroxyapatite, containing altogether 0.36% F and 0.3% Cl.

Francolite (Staffelite) — $Ca_{10}P_{5.2}C_{0.8}O_{23.2}F_{1.8}OH$. This mineral is hexagonal. The hardness if 4, the specific gravity 3.1-3.2. The indices of refraction are No = 1.629 and Ne = 1.624; the birefringence is low, 0.005-0.006; the mineral is frequently biaxial, negative, with Ng = 1.627-1.630, Np = 1.614-1.617, and Ng−Np = 0.013. Francolite forms sintery incrustations and finely porous and compact masses that are white, gray, or yellow in color. In addition to the crystalline form, the mineral is known to occur in optically amorphous varieties called collophane or collophanite.

Francolite is the chief constituent of many of the world's largest bedded deposits of geosynclinal phosphorites.

Collophane (Collophanite). A colloidal, optically amorphous phosphate of calcium, apparently representing an optically amorphous variety of francolite. The hardness if 2.0-3.5, the specific gravity 2.5-2.9. It occurs in nodular aggregates, oolites, and earthy masses; these masses are complex, ranging from friable to hard and dense. The color is variable, from white and light yellow to brown-black. The mineral appears amorphous under the polarizing microscope, but x-ray data and electron-microscopic examination show it to be a cryptocrystalline mineral. The index of refraction ranges from 1.57-1.62, but according to other authors, from 1.61 to 1.63.

Collophane is characteristic of sedimentary rocks, apparently constituting, with francolite, the main bulk of many phosphorites in the principal deposits of the bedded-geosynclinal type.

Kurskite — $Ca_{10}P_{4.8}C_{1.2}O_{22.8}F_2(OH)_{1.2}$. This mineral was first distinguished by V. N. Chirvinskii [1919], who assigned it the formula $Ca_8(PO_4)_4(CO_3)F_2$. The crystal system is hexagonal. The specific gravity is 2.9-3.0 (3.0 for pure kurskite). The color is generally gray or brown (from organic, humic, or ferruginous impurities), sometimes white or black. The index of refraction ranges from 1.59 to 1.61; the mineral is optically amorphous, cryptocrystalline, or distinctly crystalline with No = 1.597-1.610; Ne = 1.590-1.602, and with low birefringence (0.007-0.008).

Kurskite forms nodular or platform-type phosphorites, widespread within the USSR, and is found in two principal varieties: radiating (previously incorrectly called staffelite) and optically amorphous (under the electron microscopic proving to consist of microscopic crystals). The nodular growths of kurskite generally contain a large quantity of clastic quartz and glauconite grains, in addition to calcite, pyrite, iron hydroxides, and clay particles. Finally, kurskite also forms so-called phosphorite flags, which represent several generations of nodular layers, in which the phosphorite nodules have been joined together by phosphate of a later generation.

Carbonate-Apatite (Podolite or Dahllite) — $Ca_{10}P_5CO_{24}(OH)_3$. The formula for podolite was first considered to be $Ca_{10}(PO_4)_6(CO_3)$. The crystal system is hexagonal. The hardness is 4-5; the specific gravity ranges from 2.93-3.0 to 3.13. The color is white or yellow, sometimes greenish; it may be dark gray because of organic impurities. The indices of refraction are No = 1.635-1.628 (sometimes as low as 1.603) and Ne = 1.631-1.619 (sometimes as low as 1.598); No−Ne = 0.004-0.008; the mineral is uniaxial and negative.

This compound has been found in cavities in phosphorite nodules that were reworked from the Cambrian-Silurian strata of the Podolia to form the basal deposits of the Cenomanian [Chirvinskii, 1907]. As has been explained [Furman, 1954], the main bulk of these reworked spherical phosphorite nodules is composed chiefly of fluorhydroxyapatite, i.e., a mineral intermediate between fluorapatite and hydroxyapatite. Nodules of primary phosphorites in lower paleozoic rocks in Podolia consist of fluorapatite. Hydroxyapatite in which a small part of the OH has been replaced by F and Cl has been found here in the inner zones of the reworked phosphorite concretions. Carbonate-apatite was not found in the specimens studied by E. P. Furman.

The calcium phosphates in the phosphorites have recently been studied by the powder method of x-ray structural analysis. The x-ray structural pattern of the sedimentary calcium phosphates is, in some cases, identical, in other cases, similar to the lattice of fluorapatite.

Iron Phosphates

Sedimentary iron phosphates are found in some iron-ore deposits, as lenses in peat layers, and in disseminated forms; they also occur in certain sedimentary rocks (clays, sandy-silty clays, etc.) as well as in modern peat deposits. Sedimentary iron phosphates include vivianite and paravivianite, kerchenite, oxykerchenite, bosphorite, picite, beraunite, and borickite.

Vivianite — $Fe_3^{\cdot\cdot}(PO_4)_2 \cdot 8H_2O$. A hydrous phosphate of ferrous iron, having the chemical composition 43% FeO, 28.3% P_2O_5, and 28.7% H_2O. A variety of vivianite containing $Mn^{\cdot\cdot}$ and Mg, or $Mn^{\cdot\cdot}$ and Ca, is called paravivianite (for example, 2.01% MnO, 1.32% MgO, 0.48% CaO). The mineral crystallizes in the monoclinic system. The hardness is 1.5-2.0, the specific gravity 2.71-2.95 (for vivianite) and 2.60-2.68 (for paravivianite). The mineral occurs is prismatic crystals, radiating and other types of aggregates, and also in earthy masses. Unoxidized ("fresh") vivianite is very faintly colored (whitish) or colorless, but on partial oxidation in air, it becomes blue or grayish blue, dark blue to black, grayish green, dark green to greenish black; in the process of oxidation vivianite changes to kerchenite (see pp. 87). The streak of vivianite is colorless, but oxidized varieties have a bluish white or dark blue streak. Vivianite has perfect cleavage along (010). The indices of refraction are Ng = 1.620 to 1.633-1.636 (less for vivianite, more for paravivianite — Ng > 1.628), Nm = 1.605-1.604 to 1.596, Np = 1.583-1.590 to 1.578-1.570 (greater for vivianite, less for paravivianite); Ng−Np = 0.037 for vivianite and near 0.054 for paravivianite (according to L. O. Stankevich); the mineral is biaxial and optically positive.

Vivianite is distinguished by alteration (darkening) of its whitish color in air, partially oxidized varieties acquiring a characteristic light blue of grayish blue color and a corresponding streak; the hardness is low. Paravivianite is distinguished from vivianite by lower specific gravity, by greater birefringence, and by chemical composition. They both dissolve easily in hydrochloric and nitric acids.

Vivianite is a mineral of sedimentary deposits, forming in a slightly reducing environment. It is common in sedimentary iron-ore deposits (that are rich in phosphorus) and in peat bogs (where it generally occurs as an earthy variety). It is found with siderite, chamosite, and other ferrous minerals. Radiating-acicular, star-shaped, and lamellar aggregates of vivianite and kerchenite are frequently found in cavities in fossil animal bones and shells or in cavities in brown iron nodules in deposits on the Kerch and Taman peninsulas.

Kerchenite. A hydrous phosphate of ferrous iron (vivianite and paravivianite) in which part of the $Fe^{..}$ has been oxidized to $Fe^{...}$; as a result of this oxidation the whitish or colorless appearance of vivianite and paravivianite changes to light blue or grayish blue, dark blue to black-blue, grayish green, dark green to greenish black. Distinctions are made among γ-, β-, and α-kerchenites [Popov, 1910, 1929; Kantor, 1938], differing among themselves in an ever-increasing content of Fe_2O_3; in α-kerchenite, Fe_2O_3 has become more abundant than FeO.

According to S. P. Popov, β-kerchenite has the formula $5RO \cdot 2Fe_2O_3 \cdot 3P_2O_5 \cdot 23H_2O$, whereas α-kerchenite is $3FeO \cdot 3Fe_2O_3 \cdot 3P_2O_5 \cdot 21H_2O$. M. I. Kantor distinguished γ-kerchenite as the very first product of oxidation of vivianite, giving its formula as $7(Fe,Mg,Mn,Ca)O \cdot Fe_2O_3 \cdot 3P_2O_5 \cdot 23H_2O$. The final products of oxidation of vivianite are bosphorite and oxykerchenite (see below and p. 88). The chemical composition of β-kerchenite is 0.11% CaO, 0.09% MgO, 0.08% MnO, 23.4% FeO, 20.32% Fe_2O_3, 28.25% P_2O_5, and 27.38% H_2O.

The varieties γ- and β-kerchenites are light blue, dark blue to blue-black, though β-kerchenite may even be green; α-kerchenite is green, dark green, or greenish black. Oxidation occurs with no clear change in the crystalline structure; i.e., kerchenites, like vivianite and paravivianite, are monoclinic. The hardness is 1.5-3.0, the specific gravity 2.500-2.695 (less for parakerchenites, 2.5-2.6). The Kerch vivianite (paravivianite) has a specific gravity of 2.63-2.66, whereas the α-kerchenite (parakerchenite) has a specific gravity of 2.51-2.58. The indices of refraction (according to L. O. Stankevich) are Ng = 1.637-1.665, Nm = 1.604-1.660, and Np = 1.582-1.626 for β-kerchenite, Ng = 1.655-1.696, Nm = 1.633-1.682, and Np = 1.618-1.645 for α-kerchenite; the kerchenites are biaxial and negative.

According to F. V. Chukhrov and L. P. Ermilova [1956], the γ-kerchenite of Zheleznyi Rog (forming pseudomorphs after wood) has the refractive indices Ng = 1.648 and Nm = 1.616 and is biaxial, negative; but the magnesian γ-kerchenite of Kamysh-Burun (also forming pseudomorphs after wood) has the indices Ng = 1.634 and Np = 1.586, and is biaxial, negative.

According to F. V. Chukhrov and L. P. Ermilova [1956], the formulas for γ-, β-, and α-kerchenites have the following forms (it is assumed that the ferric oxide in excess of phosphoric acid is associated with the hydroxyl groups):

$$\gamma\text{- kerchenite} \quad - R^{..}_{7.79-7.06} \cdot Fe^{...}_{2.06-2.58} \cdot (OH)_{3.34-4.28} \times$$
$$\times (PO_4)_6 \cdot 20.09 - 22.15\ H_2O;$$

$$\beta\text{- kerchenite} \quad - R^{..}_{5.56-4.40} \cdot Fe^{...}_{3.68-3.94} \cdot (OH)_{2.62-4.36} \times$$
$$\times (PO_4)_6 \cdot 20.13 - 21.23\ H_2O;$$

$$\alpha\text{- kerchenite} \quad - R^{..}_{3.54-2.68} \cdot Fe^{...}_{6.00-6.48} \cdot (OH)_{5.36-8.52} \times$$
$$\times (PO_4)_6 \cdot 17.57 - 19.25\ H_2O.$$

F. V. Chukhrov and L. P. Ermilova believe that there is a gradual transition from vivianite to bosphorite and oxykerchenite and that the formulas proposed by S. P. Popov and M. I. Kantor correspond to individual points in this process.

Kerchenites (and parakerchenites) are found in a number of sedimentary iron-ore deposits rich in phosphorus (Kerch and Taman peninsulas) and in present-day peat bogs, where they are generally earthy varieties.

Oxykerchenite. A hydrous oxide of ferric iron, generally forming by complete oxidation of paravivianite or vivianite through intermediate products of oxidation (kerchenite and parakerchenite).

According to S. P. Popov [1910, 1929], the formula for oxykerchenite is $RO \cdot 4Fe_2O_3 \cdot 3P_2O_5 \cdot 21H_2O$. F. V. Chukhrov and L. P. Ermilova have data that indicate the following general formula for oxykerchenite: $R^{..}_{2.47-1.99}Fe^{...}_{8.00-8.64}(OH)_{8.56-12.56}(PO_4)_6 \cdot 14.04-16.75\ H_2O$.

A chemical analysis of oxykerchenite derived from paravivianite on the Kerch Peninsula shows 0.79% CaO, 1.22% MgO, 2.57% MnO, 41.82% Fe_2O_3, 28.04% P_2O_5, and 24.98% H_2O: a total of 99.42%. The specific gravity is 2.65.

According to L. O. Stankevich, oxykerchenite forms pseudomorphs after paravivianite. The hardness is 3.0-3.5, the specific gravity 2.45-2.58. The mineral is brown or red-brown, and has a brown-yellow streak. In thin sections it is pale yellow; it is optically isotropic or weakly anisotropic. The indices of refraction range from 1.698-1.702 to 1.712-1.716; the birefringence is very weak, near 0.002.

Oxykerchenite is found with picite and hydrogoethite.

Some authors believe that oxykerchenite is derived only from paravivianite, i.e., that it is a product of further oxidation of parakerchenite; they consider bosphorite to be an oxidation product of kerchenite.

Bosphorite. A colloidal hydrous phosphate of ferric iron, apparently being a product of complete oxidation of vivianite. The formula, according to S. P. Popov [1929] is $3Fe_2O_3 \cdot 2P_2O_5 \cdot 17H_2O$. F. V. Chikhrov and L. P. Ermilova [1956] give the following general formula for bosphorite: $R^{\cdot\cdot}_{0.74-0.40} Fe^{\cdot\cdot\cdot}_{8.78-9.26}(OH)_{9.82-10.53} \cdot (PO_4)_6 \cdot 18.63-20.76\ H_2O$. The specific gravity is 2.53. The mineral is brown or yellow. It is optically amorphous (isotropic), with variable indices of refraction: N is near 1.67 (from 1.669 to 1.675), but some grains have an index of 1.63-1.64 (according to L. O. Stankevich).

Picite — $Fe_4^{\cdot\cdot\cdot}(PO_4)_2(OH)_6 \cdot nH_2O$. A colloidal hydrous phosphate of ferric iron. F. V. Chukhrov and L. P. Ermilova [1956] adopt n = 21 for picite and n = 51 for delvauxite. An analysis of picite with a small quantity of impurities [Sidorenko, 1944] shows 47.32% Fe_2O_3, 2.39% CaO, 22.56% P_2O_5, 0.42% CO_2, 1% SiO_2, 26.03% H_2O: a total of 99.72%. According to A. V. Sidorenko, picite has the formula $2Fe_2O_3 \cdot P_2O_5 \cdot 9H_2O$. The hardness is 2-3-4, the specific gravity 2.38-2.83. The color is dark brown or brown-red, the streak yellow or red. The index of refraction is N = 1.64-1.68; the mineral is optically amorphous. It is known to occur in the iron ores of the Kerch Peninsula and in other places.

Beraunite — $Fe^{\cdot\cdot}Fe_4^{\cdot\cdot\cdot}(PO_4)_3(OH)_5 \cdot 3H_2O$. A basic phosphate of ferric and, partly, ferrous iron; it is monoclinic. The hardness is 3-4, the specific gravity 2.80-2.99-3.08. The color is reddish brown or red, the streak yellow. The mineral occurs in druses, concretions with radiating structure, and in incrustations. The indices of refraction are Ng = 1.815-1.820, Nm = 1.786, and Np = 1.775; Ng−Np = 0.040; the mineral is biaxial positive.

Beraunite is generally found in a state of complete oxidation, its approximate formula being $Fe_5^{\cdot\cdot\cdot}(PO_4)_3(OH)_6 \cdot 2-3H_2O$.

Borickite — $CaFe_4^{\cdot\cdot\cdot}(PO_4)_2(OH)_8 \cdot 3H_2O$. A colloidal hydrous phosphate of ferric iron. The hardness is 3-4, the specific gravity 2.7. It is found in dense colloform masses. The color is reddish brown. The index of refraction is N = 1.57-1.67; the mineral is optically amorphous. It dissolves in acid.

7. SULFATE AND FLUORIDE GROUP

The principal sedimentary sulfates, except for those restricted to salt lakes and lagoons, are alunite, jarosite, coquimbite, copiapite, melanterite, chalcanthite, barite, and celestite; with the sulfates we shall consider one fluoride: fluorite.

Alunite (Rock Alum) — $KAl_3[SO_4]_2[OH]_6$. The chemical composition is 11.4% K_2O, 37.0% Al_2O_3, 38.6% SO_3, and 13.0% H_2O; frequently the K_2O is replaced to approximately half its normal content by Na_2O, and the mineral is then called natroalunite, $(K, Na) Al_3(SO_4)_2(OH)_6$. The mineral crystallizes in the trigonal system. The hardness is 3.5-4.0, the specific gravity 2.6-2.8. Alunite sometimes occurs in small rhombohedral or thick tabular crystals. The color is white, rarely grayish, reddish, or yellowish. Cleavage is present along (0001). The indices of refraction are Ne = 1.592 and No = 1.572; Ne−No = 0.020; for natroalunite Ne = 1.595, No = 1.585, and Ne−No = 0.010; both are uniaxial positive. Alunite is generally found in fine-grained, earthy, sometimes fibrous masses.

Alunite is distinguished from similar white minerals by its optical properties and its chemical reactions. When it is moistened by a solution of cobalt nitrate it becomes blue (reaction for aluminum).

It is insoluble in water and hydrochloric acid; it dissolves with difficulty in concentrated sulfuric acid.

Alunite may be of hydrothermal or exogenetic origin, in the latter case being chiefly a product of weathering (concretionary forms in sands, clays, and bauxites). Sedimentary alunite is related to the action of surface sulfuric acid waters on aluminous rocks: clays, volcanic tuffs, and clayey limestones. Concretions of alunite in clays (kaolinitic and others) may be syngenetic.

Jarosite — $KFe_3^{\cdots}(SO_4)_2(OH)_6$. The chemical composition is 9.4% K_2O, 47.9% Fe_2O_3, 31.9% SO_3, and 10.8% H_2O. If the potassium is replaced by sodium, the mineral is then called sodium jarosite (natrojarosite). Jarosite crystallizes in the trigonal system. The hardness is 2.5-3.5, the specific gravity 3.15-3.30. Small rhombohedral crystals are found in druses, lining cavities. The color is yellow, ochrous yellow, or yellow-brown. The indices of refraction are No = 1.820 and Ne = 1.715; the birefringence is high (No−Ne = 0.105); the mineral is uniaxial negative. In natroalunite No = 1.832, Ne = 1.750, and No−Ne = 0.082.

Jarosite is distinguished from iron hydroxides of the hydrogoethite subgroup by chemical reactions and by rubbing between the fingers (it gives one the sensation of a greasy mass, whereas hydrogoethite is a hard, silty or sandy material). Jarosite does not dissolve in water, but it dissolves completely in hydrochloric acid (in weak acid on heating); it yields a precipitate of $BaSO_4$ in a solution of $BaCl_2$.

Jarosite is very frequently found in the oxidized zones of iron sulfide (pyrite) deposits, as concretions in normal sedimentary rocks (after pyrite or marcasite in many clays), and as scattered powdery material in clays (in Maikopian clays, for example) near the ground surface. In a water-air environment it yields iron hydroxides.

Coquimbite — $Fe_2^{\cdots}(SO_4)_3 \cdot 9H_2O$. A hydrous iron sulfate, crystallizing in the hexagonal system. The hardness is 2, the specific gravity 2.1. Small crystals on the surfaces of cavities are rhombohedral and tabular. The color of the mineral is dark yellow, yellowish white, or violet. The indices of refraction are Ne = 1.557 and No = 1.550; Ne−No = 0.007 (anomalous interference colors are common); the mineral is uniaxial positive. It is found in fine-grained aggregates or as small crystals in cavities. It dissolves in water.

Coquimbite is known to be an oxidation product of iron sulfides in arid climates, and also to occur in association with fumarole activity. In sedimentary rocks it has been noted at Samarskaya Luka, in the Alapevsk district of the Urals, and in other localities.

Copiapite — $MgFe_4^{\cdots}(SO_4)_6(OH)_2 \cdot 18H_2O$. This is one of the numerous hydrous sulfates of trivalent iron known to occur in nature. The crystal system in triclinic. The hardness is 2.5, the specific gravity 2.1-2.2. The color is generally gray-yellow, rarely reddish or greenish. The mineral occurs in earthy or scaly aggregates, incrustations, efflorescences, or hexagonal plates. The indices of refraction are Ng = 1.575-1.600, Nm = 1.528-1.550, and Np = 1.506-1.540; Ng−Np = 0.057-0.070; the mineral is biaxial, positive, and pleochroic. It dissolves in water.

Copiapite is frequently found in the oxidized zone of iron-ore (sulfide-bearing) deposits. It forms efflorescences in hot, arid climates.

Melanterite (Iron Vitriol) — $FeSO_4 \cdot 7H_2O$. A hydrous iron sulfate, crystallizing in the monoclinic system. The hardness is 2, the specific gravity 1.8-1.9. Crystals are rhombohedral, sometimes acicular. The color is most commonly light green, rarely dark gray. The indices of refraction are Ng = 1.486, Nm = 1.478, and Np = 1.471; Ng−Np = 0.015; the mineral is biaxial and positive.

Melanterite is distinguished from other easily soluble iron sulfates by color and by reaction for ferrous iron: it yields a greenish precipitate with ammonia, $Fe(OH)_2$, which is slowly oxidized in air. Melanterite dissolves easily in water. On heating it dissolves in its own water of crystallization.

It precipitates from supersaturated sulfate waters where there is a deficiency of oxygen; it forms in the clays of coal deposits (probably from pyrite) and, more frequently, in rich pyrite ores below the oxidized zone.

Chalcanthite (Copper Vitriol) — $CuSO_4 \cdot 5H_2O$. A hydrous copper sulfate in the triclinic system. The hardness is 2.5, the specific gravity 2.1-2.3. The color is light blue, medium blue, dark blue, sometimes with a greenish tinge. The indices of refraction are Ng = 1.546, Nm = 1.539, N= = 1.516; Ng−Np = 0.030; the mineral is biaxial and optically negative. It dissolves easily in water, straining the solution blue; it is characteristic that an iron needle immersed in a solution of copper vitriol becomes covered with metallic copper.

Chalcanthite is found in compact granular masses or in stalactitic sintery forms with radiating structure.

The mineral is found in oxidized zones of copper sulfide deposits in regions of arid climate.

Barite (Heavy Spar) — $BaSO_4$. Barium sulfate, commonly containing a small isomorphous admixture of Sr and Ca; when the content of strontium becomes considerable, the mineral is called barytocelestite (apparently there is an entire isomorphous series between barite and celestite, with intermediate properites). Barite crystallizes in the orthorhombic system. The hardness is 3.0-3.5, the specific gravity 4.3-4.7. Crystals are tabular, less commonly prismatic. They are mostly white or gray, yellow or brown (from impurities of iron hydroxides), dark gray or black (from carbonaceous material), rarely some other color; minerals free of impurities are transparent and colorless. Cleavage is perfect along (001) and (110), imperfect along (010). The indices of refraction are Ng = 1.648, Nm = 1.637, and Np = 1.636; Ng−Np = 0.012; the mineral is biaxial, positive. Elongation is positive; prismatic grains have parallel extinction. Barite is found in granular masses and radiating aggregates.

Barite is distinguished from other sulfates by its high specific gravity. It is distinguished by its perfect cleavage in one direction, and by its insolubility in hydrochloric acid even after heating (in contrast to carbonates). Thin sections exhibit relief, shagreen surface, and cleavage cracks. Barite is very similar to celestite, from which it may be distinguished in thin section by somewhat greater birefringence, in immersion liquids by somewhat higher indices of refraction, and in hand specimens or powder by chemical reactions. Barite decrepitates in front of the blow pipe, fusing only in thin slivers, and in so doing coloring the flame yellow-green, the characteristic color of barium (celestite fuses into a white bead and colors the flame a carmine-red). Powdered barite will dissolve slowly in concentrated sulfuric acid.

Barite is generally authigenic in sedimentary rocks; it is frequently found in small quantities, chiefly in concretions, but also occurs as cement in sandstones and siltstones, in clay rocks, sedimentary manganese and iron ores, and occasionally in carbonate rocks (limestones and dolomites); it sometimes forms veinlets in limestones. Barite is a rather common mineral in hydrothermal deposits.

Celestite − $SrSO_4$. Strontium sulfate, frequently containing isomorphous admixtures of Ca and Ba; there is apparently a continuous isomorphous series between celestite and barite. When the content of barium is large in the mineral, it is called barytocelestite. Celestite crystallizes in the orthorhombic system. The hardness is 3.35, the specific gravity 3.9-4.0. Crystals are tabular or prismatic. The color is light bluish white or light bluish gray, rarely exhibiting reddish or yellowish tints; the mineral is sometimes colorless. Cleavage is perfect along (001) and (110), imperfect along (010). The indices of refraction are Ng = 1.631, Nm = 1.624, and Np = 1.622; Ng−Np = 0.009; the mineral is biaxial, positive. In barytocelestite the optical and other properties are intermediate between celestite and barite: an increase in the barium content is accompanied by an increase in specific gravity and in indices of refraction (Np from 1.624 to 1.636).

Celestite is distinguished from anhydrite by higher specific gravity, from carbonates by insolubility in hydrochloric acid. It is distinguished from the similar barite by somewhat lower birefringence (viewed in thin section), by somewhat lower indices of refraction (measured in immersion liquids) (in each section Np' = 1.622-1.624 in celestite as against Np' = 1.636-1.637 in barite, intermediate values indicating barytocelestite as indicated on p. 89), and by a test for strontium (in hand specimens and powders): before the blow pipe celestite fuses into a white bead, the flame becoming a deep carmine-red (barium colors the flame yellow-green). Celestite dissolves in strong sulfuric acid.

Celestite is almost always authigenic in sedimentary deposits (a chemical precipitate) and is associated with lagoonal deposits of high salinity: dolomites, gypseous clays, marls, and dolomitic limestones, sulfate-dolomite rocks, gypsum rocks, and anhydrite rocks. The mineral is rarely found in hydrothermal veins and in amygdules in volcanic rocks.

The origin of the sedimentary celestite in the Permian rocks of Tataria has been discussed in a monograph by L. M. Miropol'skii [1926]. Celestite is found in this locality in various forms in dolomites, gypsum rocks, and chert.

Most recently celestite has been described from many localities in European USSR (chiefly in Permian rocks, parly in Devonian deposits) and also from separate localities in Asiatic USSR.

Fluorite (Fluorspar) − CaF_2. The chemical composition is 51.2% Ca and 48.8% F. The mineral forms in the isometric system. The hardness is 4, the specific gravity 3.0-3.2, 3.15-3.18-3.20 in pure varieties. Crystals are cubic, or exhibit combinations of the cube and the octahedron or dodecahedron. The color is generally violet, light blue, green, or yellow, sometimes violet-black, or, in contrast, colorless. Perfect octahedral cleavage is present. The index of refraction is distinctly low (N = 1.433-1.434); the mineral is isotropic, but sometimes shows anomalous polarization.

Fluorite is recognized by its cubic form, octahedral cleavage, isotropic behavior, and low index of refraction; in thin sections it is colorless or faintly violet. It decomposes completely only in strong sulfuric acid, with the elimination of HF. The effect of nitric acid or hydrochloric acid is much weaker.

90

Hydrothermal, as well as sedimentary, fluorite has been recognized. In sedimentary deposits flurotie may be either authigenic or fragmental. CaF_2 is one of the first compounds to separate out during evaporation of the water in saline lagoons and of salt water in general. The accumulation of sedimentary fluorite is therefore associated with pelitomorphic chemical dolomites of Paleozoic age, and also with gypsum and anhydrite; fluorite sometimes precipitates in very small quantities in marine limestones. At times sedimentary fluorite has formed not because of high mineralization of the water but because of a sharp increase in the concentration of fluorine combinations in water derived from fluorine-rich rocks along the shore or from such rocks forming individual elevations on the land surface near the basin. Fluorite in the cement of certain red sandstones in the Vereyan horizon of the Moscow Basin probably formed chiefly in this manner.

The formation of earthy sedimentary fluorite, called ratofkite, is due to epigenetic processes. The name derives from Ratovka Creek at Veren in the Moscow district, where it was first described in dolomitic limestones and dolomites of Carboniferous age. Ratofkite has been recognized in the upper Volga region [Pustovalov, 1937], in the Donets Basin, and at other localities in the USSR. Ratofkite has formed by the redeposition of CaF_2 and by its reprecipitation in amorphous, colloidal form. Under the microscope ratofkite proves to consist of very fine (0.03–0.05 mm) rounded bodies, similar to opaline globules, but distinguished from them by their faint violet color.

Sedimentary fluorite has been found in various parts of the Soviet Union: in the Moscow Basin (Carboniferous), along the Northern Dvina (Permian deposits), in the upper Paleozoic rocks of the Ural-Volga region (chiefly in Lower Permian strata), in the Ural-Emba region (in the same strata), in Lower Cambrian rocks of the Aldan district of the Yakut ASSR, and elsewhere.

8. SOLUBLE SALT GROUP (LACUSTRINE AND LAGOONAL SALTS)

We shall consider here only the most common natural soluble salts, particularly, gypsum, anhydrite, halite, sylvite, carnallite, kainite, polyhalite, glauberite, kieserite, epsomite, mirabilite, thenardite, and nitratine.

Gypsum - $CaSO_4 \cdot 2H_2O$

The chemical composition is 32.5% CaO, 46.6% SO_3, and 20.9% H_2O. The mineral is monoclinic. The hardness is 1.5–2.0 (it may be scratched by the fingernail), the specific gravity 2.2–2.4 (normally 2.3). Gypsum crystals are chiefly tabular, sometimes prismatic. The color is white, sometimes water-clear and transparent; on the other hand, gypsum may be stained various colors by pigmenting impurities. Cleavage is perfect along (010), distinct along (100) and (111). The indices of refraction are $Ng = 1.529–1.5305$, $Nm = 1.523$, and $Np = 1.520–1.521$; $Ng-Np = 0.009–0.0095$; the mineral is biaxial and positive, and the elongation of fibrous individuals is negative. Gypsum is found in compact granular masses, also as crystals in cavities and as normal fibrous varieties (selenite) in fractures.

Gypsum is distinguished by the low hardness (scratched by the fingernail) and by the perfect cleavage along (010) in crystals. It does not give off bubbles of CO_2 when moistened with hydrochloric acid; in water acidified with sulfuric acid (up to a certain limit—75 g/liter), it dissolves better than inpure water; powdered gypsum dissolves in 5 or 10% hydrochloric acid (on boiling), in nitric acid, in solutions of $(NH_4)_2SO_4$, in concentrated $(NH_4)Cl$, and in other reagents. The solubility of gypsum in water at 20° is 0.224% by weight.

Four phases are known in the system $CaSO_4 \cdot H_2O$: $CaSO_4 \cdot 2H_2O$ (gypsum), $2CaSO_4 \cdot H_2O$ or $CaSO_4 \cdot \frac{1}{2}H_2O$ (bassanite), γ-$CaSO_4$ ("soluble anhydrite"), and β-$CaSO_4$ (anhydrite). In water, gypsum changes with time to anhydrite at temperatures above 42°; at lower temperatures the reverse process is effective. On heating, gypsum changes to bassanite (in air, slowly at 70° and rapidly at 90–130°; in water, at 98°). If, in preparing a thin section, the gypsum is heated above these temperatures, it is partly or completely changed to bassanite, which will show distinct interference colors and brownish tones in the thin section.

To reach the saturation point of $CaSO_4$ (as gypsum) in solution, the mineralization of normal sea water at 30° would have to be increased to 3.5 times its value; if the mineralization were increased to five times its value, gypsum would separate out, but on further increase in salinity, anhydrite would become the stable phase. At 25° $CaSO_4$ will, for the most part, separate out over a long period of time as gypsum during evaporation of sea water; under such circumstances, anhydrite begins to precipitate from solution only when the solution reaches the saturation point for NaCl or approaches that point; i.e., anhydrite separates out with the first part of halite accumulation.

In a system of pure water and $CaSO_4$, anhydrite is the stable phase above 42-66°, but in sea water, where there is an abundance of NaCl in solution and where $MgCl_2$ is present, anhydrite may form at a temperature of 30° or even 25° if the mineralization is sufficiently high (see below).

Gypsum forms in nature by various methods: a) chiefly by precipitation from solution in saline lagoons or salt lakes (when the temperature and the concentration of soluble salts are high, anhydrite precipitates with the gypsum); b) to a considerable extent during epigenetic hydration of anhydrite in sedimentary deposits of saline lagoons by the action of surface waters (to depths of 100-150 m below the ground surface); c) in the weathering zones at a number of rock-salt deposits and other salt concentrations by hydration of the contained anhydrite and partly by decomposition of certain salts by water, for example, polyhalite and glauberite ("gypsum cap"); d) in the weathering zone in deserts and semideserts; e) during oxidation of pyrite or other iron sulfides in calcareous or lime-bearing rocks (an association of gypsum and iron hydroxides develops); f) rarely as a hydrothermal mineral in epithermal sulfide deposits, forming at low pressures and temperatures.

Gypsum may change to anhydrite in lithified rocks by a sharp increase in the mineralization of the ground water.

Anhydrite - $CaSO_4$

Anhydrous calcium sulfate with the chemical composition 41.2% CaO and 58.8% SO_3; the system is ortho-rhombic. The hardness is 3.0-3.5, the specific gravity 2.80-3.00. Crystals of anhydrite are thick-tabular or prismatic. The color is white, commonly with a light blue or grayish tinge, or blue; colorless crystals may sometimes be found. Cleavage is perfect along (001) and (010), moderate along (100). The indices of refraction are Ng = 1.613-1.614, Nm = 1.575-1.576, and Np = 1.569-1.571; Ng-Np = 0.043-0.044; the mineral is biaxial and positive, with parallel extinction and negative elongation.

Anhydrite is best distinguished from the similar polyhalite and kieserite by the fact that it is almost insoluble in water; in immersion liquids it is distinguished from polyhalite by a slightly higher index and from kieserite by a distinctly higher index of refraction. Anhydrite dissolves slightly in hydrochloric acid, which distinguishes it from carbonates; in thin section it shows lower (bright) interference colors than carbonates. Anhydrite is distinguished from gypsum by greater hardness (cannot be scratched by the fingernail), from barite by lower specific gravity, and, in thin section, from both by greater birefringence (clear interference colors), and by right-angle cleavage in three directions.

Anhydrite is a typical chemical precipitate in saline lagoons, but it may form epigenetically by a sharp increase in the mineralization of ground waters. Fine-crystalline and microcrystalline anhydrite quickly precipitates from solution; but distinctly crystalline anhydrite forms slowly, frequently metasomatically[Teodorovich, 1942[1],1950]. To attain the saturation point of anhydrite in solution, the mineralization of normal sea water at 30° must be increased to five times its value, and at 25° to almost 10 times its value (approximately to the point where NaCl first begins to precipitate from the solution).

Anhydrite is generally found (with gypsum, rock salt, and dolomite) in deposits of saline lagoons chiefly of Paleozoic age, having formed in beds, seams, layers, lenses, or disseminations. In anhydrite masses that have formed by evaporation of normal sea water (lagoons), the principal carbonate (impurity) is generally dolomite, especially when the partial pressure of carbon dioxide in the atmosphere is appreciable or high; occasionally, when the salinity approaches the saturation point for NaCl, the chief carbonate may be magnesite (or hydro-magnesite), sometimes calcite. Anhydrite is abundant in primary spotty anhydrite-dolomite rocks, which form by the simultaneous precipitation of dolomite and anhydrite and which are widespread among Paleozoic lagoonal deposits [Teodorovich, 1942[1], 1946[2], 1950, 1955].

Anhydrite also forms epigenetically, by metasomatism or by slow precipitation from concentrated calcium chloride and similar deep ground waters, in which its solubility has been sharply decreased. This may be the origin of anhydrite in many organic limestones, in the cement of certain sandstones, and in other occurrences. Finally, one may sometimes find anhydrite of hydrothermal or contact-metasomatic origin.

When cold surface waters act on anhydrite, the mineral combines with the water, changing to gypsum ($CaSO_4 \cdot 2H_2O$) and increasing noticeably in volume. This property of anhydrite may lead to disturbances in the attitude of adjacent beds.

Halite (Rock Salt) – NaCl

An isometric mineral. The hardness is 2.0–2.5, the specific gravity 2.1–2.2. Crystals are cubic in form. Pure halite is colorless and transparent or white; impurities may stain it various colors – gray, yellow, red, brown, or black. A characteristic deep blue color is sometimes found in bands or spots, especially in zones where deformation has been strong. Cubic cleavage is perfect. The index of refraction is 1.544; the mineral is isotropic.

Halite dissolves easily in water (giving the solution a salty taste); thin sections should therefore be prepared in glycerine. The mineral may also be recognized by its low hardness.

Halite content is determined chemically from a solution of $AgNO_3$ acidified with nitric acid by the precipitation of a white clotted precipitate of AgCl. When the content of NaCl (more properly Cl') is small in a rock, the material may be detected in the following way: 1) take a small sample of powdered rock (0.5 g, for example) and shake with distilled water (for surer results the contents may be heated slightly); 2) filter off the pure solution; 3) introduce 1-2 drops of nitric acid to obtain an acid environment; 4) add a solution of $AgNO_3$ by drops. When the content of NaCl is small, the following results are obtained, according to the degree of concentration of Cl': a) extremely weak opalescence, b) very weak opalescence, c) weak opalescence, d) moderate opalescence, e) strong opalescence, f) slightly turbid solution, g) strongly turbid solution and a precipitate, the quantity of which may vary.

Halite is found in sedimentary rocks in dense coarsely crystalline masses, forming independent beds, lenses, plugs, etc., or occurring in a disseminated form. It is found on the floors of present-day salt lakes and lagoons, precipitated in unconsolidated crystalline aggregates.

Halite is a typical chemical precipitate in evaporating, sufficiently mineralized waters in strongly saline lagoons or salt lakes, precipitating after the separation out of anhydrite-dolomite rocks and anhydrite (gypsum). In order for halite to precipitate from present-day marine (oceanic) waters, the volume would have to be decreased to approximately one-tenth its present value. Halite precipitates at first alone, and later (when mineralization is high) with sylvite and other easily dissolved salts.

Sylvite – KCl

The chemical composition is 52.5% K and 47.5% Cl. Sylvite commonly contains mechanical admixtures of NaCl and Fe_2O_3. The crystal system is isometric. The hardness is 1.5–2.0, the specific gravity 1.97–1.99. Crystals have a cubic aspect. The mineral is colorless, white, bluish, rose-colored, or bright red (from inclusions of Fe_2O_3). The index of refraction is N = 1.490; the mineral is isotropic. It is found chiefly in compact granular masses. It is much less widespread in nature than halite.

Sylvite dissolves in water. The taste is bitter-salty. In thin sections, prepared in glycerine, sylvite is distinguished from halite by negative relief. In a solution of $AgNO_3$ acidified with nitric acid, sylvite, like halite, gives a white precipitate of AgCl.

And, again like halite, sylvite forms as a chemical precipitate from the waters of markedly saline lagoons and salt lakes. It is not a common mineral, since it begins to precipitate from brines very late (it requires a reduction in volume of the original oceanic waters to 1–1.5%) and is therefore found in the uppermost parts of salt deposits; it is sometimes found as a decomposition product of carnallite ($KCl \cdot MgCl_2 \cdot 6H_2O$). It occurs with halite, forming the rocks called sylvinites. During epigenesis sylvite may be replaced by polyhalite (in sulfate deposits with $MgSO_4$).

Sylvite is known to occur in a number of ancient salt deposits and in present-day potassium lakes near the sea. It is ordinarily present in three groups of saline rocks: halite rocks (rock salt), sylvite-halite rocks (or sylvinites), and halite-carnallite rocks.

Carnallite – KCl · MgCl$_2$ · 6H$_2$O

The chemical composition is 14.1% K, 8.7% Mg, 38.3% Cl, and 38.9% H_2O; the crystal system is orthorhombic. The hardness is 2-3, the specific gravity 1.6. It is normally found in compact granular masses of a rose or red color (from finely disseminated admixtures of Fe_2O_3); it is occasionally stained brown or yellow (admixtures of iron hydroxides); pure varieties are colorless. The indices of refraction are Ng = 1.494, Nm = 1.475, and Np = 1.466; Ng-Np = 0.028; the mineral is biaxial, positive; cleavage is absent. The taste is sharply bitter-salty. Carnallite is extremely hygroscopic: in air it deliquesces, decomposing to KCl and $MgCl_2 \cdot 6H_2O$. On dissolving in water, carnallite gives off a cracking sound, and when pierced by a sharp nail it emits a sharp creak (gas bubbles are set free).

Carnallite is a typical chemical precipitate from the evaporation of markedly saline waters. It is found with halite and sylvite, the three separating out one after the other during the drying up of brines in saline lagoons and salt lakes near the sea.

Kainite - $KMg(SO_4)Cl \cdot 3H_2O$ or $KCl \cdot MgSO_4 \cdot 3H_2O$

A monoclinic mineral. The hardness is 2, the specific gravity 2.1. The mineral is yellowish or grayish white, sometimes red or colorless. Cleavage is perfect along (001). The indices of refraction are Ng = 1.516, Nm = 1.505, and Np = 1.494; Ng-Np = 0.022; the mineral is biaxial and negative. It is generally found in dense granular masses.

Kainite dissolves easily in water; the taste is bitter-salty. In contrast to carnallite it is not hygroscopic; it is distinguished from that mineral also by somewhat higher indices of refraction.

It is known to occur in potassium deposits, where it is found with carnallite, kieserite, halite, and similar minerals. During weathering it changes to epsomite, polyhalite, and other minerals.

Polyhalite - $K_2MgCa_2(SO_4)_4 \cdot 2H_2O$

This mineral is triclinic. The hardness is 2.5-3.0, the specific gravity 2.72-2.78. The color is gray, white with grayish or yellowish tones, rose-colored, or brick-red. Cleavage is present along (100). The indices of refraction are Ng = 1.567, Nm = 1.562, and Np = 1.548; Ng-Np = 0.019. The mineral is found in dense, fibrous, and columnar aggregates.

Water decomposes polyhalite: it leaches K and Mg salts from it, and a precipitate, corresponding in composition to gypsum, accumulates.

Polyhalite is a mineral of salt deposits, crystallizing from brines rich in K, Mg, and Ca, at temperatures from 0 to 80°. It is widespread among potassium sulfate deposits. It may also form in place of sylvite (see above), kainite ($KCl \cdot MgSO_4 \cdot 3H_2O$), and several other saline minerals containing K.

Glauberite - $Na_2Ca(SO_4)_2$ or $Na_2SO_4 \cdot CaSO_4$

A monoclinic mineral. The hardness is 2.5-3.0, the specific gravity 2.7-2.83; Crystals are tabular or prismatic. The color is light yellow, gray, brown, rose, or red (from admixtures of finely disseminated oxides or hydroxides of Fe and other elements); pure glauberite is colorless. Cleavage is perfect along (001). The indices of refraction are Ng = 1.536, Nm = 1.535, and Np = 1.515; Ng-Np = 0.021; the mineral is biaxial and optically negative.

In thin sections glauberite is distinguished from gypsum by higher birefringence. In immersions, epsomite and mirabilite are distinguished from glauberite by lower indices of refraction.

Glauberite decomposes in water, gypsum and mirabilite forming, but if the quantity of water is large, only gypsum remains in the sediment.

Glauberite is a mineral of salt deposits, particularly primary chemical precipitates from the brines of saline basins. It is unstable in the zone of weathering.

Kieserite - $MgSO_4 \cdot H_2O$

A monoclinic mineral. The hardness is 3.5, the specific gravity 2.57. Kieserite is generally turbid, or white with yellowish tones. The indices of refraction are Ng = 1.584, Nm = 1.533, and Np = 1.520; Ng-Np = 0.064; the mineral is biaxial and positive, and has an extinction angle (c $\wedge$ Ng) of 76°. It dissolves slowly in water.

Kieserite is distinguished from epsomite by higher refractive indices and birefringence, from anhydrite, on the other hand, by lower indices of refraction; it may be distinguished from polyhalite in an immersion liquid of N = 1.567.

94

Kieserite is a mineral of salt deposits, normally being a dehydration product of epsomite ($MgSO_4 \cdot 7H_2O$) or of hexahydrite ($MgSO_4 \cdot 6H_2O$). It rarely crystallizes from brines; when it does it is among the late minerals. It gradually changes to epsomite in moist air.

Epsomite (Bitter Salt) - $MgSO_4 \cdot 7H_2O$

This mineral crystallizes in the orthorhombic system. The hardness is 2.0-2.5, the specific gravity 1.68-1.75 (commonly near 1.70). Crystals are prismatic or acicular. The mineral is white or colorless, transparent. Cleavage is perfect along (010). The indices of refraction are Ng = 1.461, Nm = 1.455, and Np = 1.433; Ng-Np = 0.028; the mineral is biaxial, negative. It dissolves easily in water. It is distinguished by a bitter-salty taste. It may be distinguished from other hydrous magnesium sulfates or more complex sulfates by chemical analysis.

Epsomite is known to occur in several salt deposits, but is most frequently found in present-day magnesium sulfate saline deposits in salt lakes, where it generally precipitates with halite. It is also known as an efflorescence in sedimentary rocks (in fractures or cavities, or on exposed surfaces of these rocks, under overhanging projections); in such occurrences epsomite is commonly associated with processes of dedolomitization. Last, hydrothermal (very low temperature) epsomite has been found. The mineral is unknown in Paleozoic salt deposits (it has changed to kieserite).

Mirabilite (Glauber Salt) - $Na_2SO_4 \cdot 10H_2O$

A sodium sulfate with abundant water, the chemical composition being 19.3% Na_2O, 24.8% SO_3, and 55.9% H_2O; the crystal system is monoclinic. The hardness is 1.5-2.0, the specific gravity 1.48-1.49. Crystals form stubby prisms. The mineral is colorless and transparent to white and turbid. Cleavage is present along (100). The indices of refraction are very low (Ng = 1.398, Nm = 1.396, and Np = 1.394); Ng-Np = 0.004; the mineral is biaxial and negative, and c $\wedge$ Ng = 31°. It generally occurs in compact granular aggregates.

Mirabilite dissolves very easily in water, is salty, bitter and cooling to the taste. It is distinguished from other easily dissolved hydrous sulfates and chlorides by its low specific gravity, its low indices of refraction, and its low birefringence; when acted on by hydrochloric acid it does not give off CO_2, a fact that distinguishes it from natron. In dry air mirabilite changes into a white powdery substance—thenardite (Na_2SO_4).

Mirabilite precipitates from waters in salt lakes saturated with sodium and sulfate ions at temperatures below 33° (at temperatures above 33° anhydrous sodium sulfate, thenardite, forms under these conditions) or it forms during marked lowering of the temperatures in autumn and winter (for example, in winter in Kara-Bogaz-Gol before 1939, when the water temperature was below 6°). If NaCl is present in the solution, thenardite precipitates with mirabilite during evaporation, beginning at somewhat lower temperatures, 33-32°.

Thenardite - Na_2SO_4

Anhydrous sodium sulfate, crystallizing in the orthorhombic system. The hardness is 2-3, the specific gravity 2.68-2.69. Crystals are dipyramidal. The mineral is normally colorless and transparent; it sometimes has a reddish tinge. It is frequently found in druses or granular aggregates. The indices of refraction are Ng = = 1.484-1.485, Nm = 1.474-1.477, and Np = 1.464-1.471; Ng-Np = 0.021-0.013; the mineral is biaxial and positive. It dissolves readily in water.

Thenardite precipitates from supersaturated solutions only at temperatures above 32.5°, although when NaCl is present it separates out at somewhat lower temperatures; in general mirabilite forms at lower temperatures.

Thenardite, like mirabilite, precipitates during evaporation of salt lakes; it also forms by dehydration of mirabilite; for example, during the summer (before 1939) on the floor of Kara-Bogaz-Gol Gulf, mirabilite was dissolved, but, being precipitated along the shore, it was changed to thenardite in the upper parts of the layers.

Nitratine (Soda Niter, Chilean Niter) - $NaNO_3$

A trigonal mineral. The hardness is 1.5-2.0, the specific gravity 2.2-2.3. The color is white, gray, reddish brown, or yellow. Cleavage is perfect along the rhombohedron (1011). The indices of refraction are

No = 1.585 and Ne = 1.337; No-Ne = 0.248; the mineral is uniaxial and optically negative. It dissolves readily in water. The taste is slightly salty and cooling.

Nitratine forms in hot, arid regions devoid of vegetation by biochemical oxidation of nitrogen-bearing organic material (from bird and animal excrement), nitrobacteria, micro-algae,etc. On rare occasions this occurrence includes niter in depressions or hollows, where either saline niter soils or continuous deposits of niter gradually accumulate. Nitratine is found with gypsum, mirabilite, and halite. It is also known from individual lakes with bottom precipitates of nitratine (Doronino Lake in the Transbaikal region) and glauber salt ($Na_2SO_4 \cdot 10H_2O$).

9. NATIVE ELEMENT GROUP

Of the native elements we shall discuss native sulfur first of all, since it represents the greatest interest in the petroleum industry, and after this native copper and iron.

Native Sulfur-S

In nature the stable form is an orthorhombic polymorphous modification: α-sulfur or, simply, sulfur. The hardness is 1-2, the specific gravity 2.05-2.08. Crystals are generally pyramidal (dipyramidal), or they form truncated pyramids (dipyramidal). The color is generally one of various tones of yellow, but sometimes brown or black (from organic impurities). Powdered sulfur is pale yellow. The luster on crystal faces is adamantine, but on fractures it is greasy. Cleavage is imperfect along (001), (110), and (111). In immersions the color is light yellow. The indices of refraction are Ng = 2.240-2.245, Nm = 2.038, and Np = 1.951-1.958; Ng-Np = 0.287; the mineral is biaxial and positive. It occurs in disseminations or in compact, sometimes earthy masses.

Sulfur is recognized by its characteristic color, by its low hardness, its greasy luster on fractures; in immersions it is identified by its color and high refractive indices, in thin sections by its marked relief and shagreen surface, and also by high-order white interference colors. Before the blow pipe, in a burner, or in a match flame, sulfur fuses and burns with a blue flame, giving off a characteristic odor (evolution of SO_2). It dissolves easily in kerosene, turpentine, and carbon disulfide, * but it does not decompose in hydrochloric or sulfuric acids. Strong nitric acid and aqua regia change sulfur to sulfuric acid. Electrical and thermal conductivity are very weak, sulfur therefore being used as an insulator.

Most native sulfur has formed by biochemical means in sedimentary (generally carbonate) deposits containing gypsum, solid and liquid bitumens, and petroleum gases. Some sedimentary sulfur is apparently syngenetic, and other is clearly epigenetic. Epigenetic sulfur may form in various ways: 1) by reduction of the sulfate ion during its migration upward or horizontally, for example, to oil-gas deposits at the boundary of the oxidation-reduction zone; 2) by oxidation of hydrogen sulfide, rising from below, to native sulfur, at approximately the same transitional interface; and 3) by the invasion of hydrogen sulfide waters into a deposit of gypsum or anhydrite. In addition, native sulfur is also known to occur in association with volcanic activity and, lastly, to form from sulfides in the lower parts of the oxidation zone of ore deposits.

Native Copper-Cu

Generally chemically pure, crystallizing in the isometric system. The hardness is 2.5-3.0, the specific gravity 8.5-8.9. Well-formed crystals are rare; they are found in twinned intergrowths, sometimes in crystalline dendrites. The color is copper-red, frequently tarnished. The luster is typically metallic. In reflected light in polished sections, copper is a bright rose-color and isotropic.

Native copper is found in rock fractures as aggregates (irregularly tabular dendrites) of exogenetic origin, or it forms small grains in rocks, incrustations, and disseminations; in oxidized zones of copper deposits it sometimes forms compact masses.

Native copper is characterized by color and luster, by its high specific gravity, malleability, hackly fracture, and high electrical conductivity. In dilute nitric acid copper dissolves readily, the solution being colored green; in hydrochloric acid it dissolves with difficulty, with the formation of copper chloride. Aqueous solutions in ammonia have a characteristic blue color.

---

*Amorphous modifications of sulfur, found as impurities in crystalline native sulfur, are insoluble in carbon disulfide.

It is of interest that metallic iron has the property of precipitating copper from sulfate solutions; iron filings are therefore used to obtain copper in the corresponding commercial manufacture.

To produce native copper, reducing conditions are necessary, an environment most common in the lower parts of the oxidation zones of copper sulfide deposits; native copper is also known from typical hydrothermal deposits or in hydrothermally altered basic magmatic rocks and in normal sedimentary rocks, principally sandstones.

In the lower parts of the oxidation zones of copper sulfide deposits, native copper occurs with cuprite, malachite, and sometimes with chalcocite and other copper minerals.

In sandstones and zones of other sedimentary rocks, native copper may be found as cement or in irregular concretions, occasionally in combination with cuprite, malachite, and azurite.

L. M. Miropol'skii, in his work on the copper ores of the Permian beds of Tataria [1938], noted the presence of native copper at only two deposits—at the villages of Krasnovidovo and Antonovka—where it occurs in gypsum rocks in the Tatarian series. At these localities native copper is observed with segregations of brochantite ($CuSO_4 \cdot 3Cu(OH)_2$) and chalcocite, but sometimes without these minerals, forming disseminated segregations; it is common to find oxidation products on the surface of the copper—cuprite and malachite. L. M. Miropol'skii associated the formation of native copper with the reduction of brochantite.

In cupriferous sandstones of the Dzhezkazgan deposit, D. G. Sapozhnikov [1948] and F. V. Chukhrov have noted native copper in a number of supergene minerals of very wide distribution. V. A. Polyanin and I. N. Gorizontova [1939] pointed out native copper in a deposit of copper ore in the Kirov district, where it occurs in microscopic segregations in dark gray Permian clays.

Native copper oxidizes in an oxygen environment (in air), becoming covered with copper oxides, but in an oxygen-rich aqueous environment it becomes coated with a crust of hydrous copper carbonates—malachite and azurite.

Native copper, in turn, may replace cuprite, chalcocite, or woody fragments, forming pseudomorphs after them.

Native Iron or α-Fe (Ferrite)

Native iron crystallizes in the isometric system. α-Fe is the low-temperature modification of iron, in contrast to γ-Fe, which is the high-temperature modification (above 906°), also crystallizing in the isometric system. The hardness is 4-5, the specific gravity 7.0-7.9. The color is steel gray; the steak also steel gray. The luster is distinctly metallic (on fresh fracture). In polished sections the color is metallic white. Native iron is generally found in extremely small grains of irregular shape, but it sometimes occurs in larger accumulations.

Distinctive features of α-Fe are its malleability and its pronounced magnetism. In air native iron oxidizes readily; it dissolves in nitric acid.

Meteoric, telluric (associated chiefly with basic and ultrabasic magmatic rocks), and exogenetic occurrences of native iron are known. The last is of special interest to us, found in "siliceous schists," i.e., foliated silicites (on Borneo), in the cements of certain fragmental rocks of Devonian age in Timan, in the products of coal-bed fires, in peat bogs, and being found as a product of reduction of organic material in combinations with oxygen (for example, brown iron ores) and of soluble iron salts in the upper parts of the lithosphere. In some cases native iron is found as incrustations on plant tissue.

Chapter IV

BASIC CONCEPTS OF COLLOIDAL CHEMISTRY
IN RELATION TO MINERAL FORMATION

Colloidal systems are of great significance for almost all phenomena associated with chemical and biogenic sediment formation and with all that happens to the sediments (weathering of parent rock, transport, deposition, and diagenesis); because of this it is advisable to devote a special discussion to colloids in this composite paper. Among the authigenic minerals in sedimentary rocks there are both typical natural colloidal as well as metacolloidal minerals; furthermore, a number of minerals may rapidly pass through the colloidal state during sedimentation.

In his work on an "Introduction to True Physical Chemistry" (1747-1752), M. V. Lomonosov recognized two types of segregations of soluble substances: crystallization and coagulation. He distinguished colloidal solutions by the term juice (sok).

The English scientist Graham (1861), in studying the rate of diffusion of various substances through animal membranes, distinguished two basic groups of substances: 1) those diffusing quickly through the membranes and crystallizing readily, called crystalloids; and 2) those not permeating through the membranes and not crystallizing, called colloids, i.e., "like glue."

Substances referred to the crystalloid group are sugar, salt, gypsum, and other substances that form true solutions on dissolving. Substances called colloids include glue, silicic acid, sulfur, and others.

In 1869 I. G. Borshchov, in his work "On the Properties and Partial Structure of Certain Colloidal Substances Participating in the Formation of Plant and Animal Organisms," concluded, in contrast to Graham, that colloidal particles are themselves crystalline. It was later discovered that a single substance may, depending on the conditions, have either the properties of a crystalloid or of a colloid. Even such a typical crystalloid as common salt makes a colloidal solution in benzene. On the other hand, soap, which forms a colloidal solution in water, manifests the properties of a crystalloid in an alcohol solution. It is thus more nearly correct to distinguish crystalloidal and colloidal states of a substance.

Mineralogists long ago began to recognize the importance of colloids in the formation of minerals, and, in particular, in the transfer of substances by natural waters. V. I. Vernadskii, in a paper in 1925, definitely pointed out the important role of colloids in most types of surface waters. The importance of colloidal material in present-day natural waters and in the processes of sedimentation has been shown by S. A. Shukarev and T. A. Tolmacheva [1930], M. A. Kulakov [1938], and by many other hydrochemists. In 1936 F. V. Chukhrov prepared a special theoretical paper, "Colloids in the Earth's Crust." The great role of colloids in sedimentation was emphasized by L. V. Pustovalov in his "Petrography of Sedimentary Rocks" [1940].

Recently F. V. Chukhrov [1955] wrote a voluminous treatise on colloids of the earth's crust, taking into account new data and describing the role of colloids in the formation of minerals during weathering, subaqueous cementation, and the development of hydrothermal deposits; this work also gives descriptions of colloidal and metacolloidal minerals.

Colloids are disperse systems, consisting of a dispersing medium and a dispersed phase (solid, liquid, or gas) and occupying an intermediate position between true solutions (molecular-ionic disperse systems) and suspensions and emulsions (coarsely disperse systems). All disperse systems are provisionally separated into the following groups, according to the size of particle in the dispersed phase:

a) Coarsely disperse systems, in which the particle dimensions in the disperse phase exceed 0.1μ (1μ according to some authors); if the disperse phase consists of solid particles, we speak of suspensions; if the disperse phase consists of very small particles of a liquid, then we speak of emulsions. Clay suspended in water is an example of a suspension; very fine particles of oil in water form an emulsion, etc. Particles in coarsely disperse systems are visible under an ordinary microscope.

b) Colloids, or colloidal systems, a disperse systems with particle dimensions of the disperse phase of 0.1μ to $1\ m\mu$, i.e., from 10^{-5} to 10^{-7} cm (according to some authors from 1μ to $1\ m\mu$, i.e., from 10^{-4} to 10^{-7} cm); colloidal particles pass through the pores of a paper filter, but do not pass through ultramicroscopically porous animal or plant membranes. The particles in colloidal systems are observed by means of an ultramicroscope or electron microscope. The settling of particles with $d < 0.1\mu$ under the influence of gravity is practically balanced by the diffusion-brownian movement.

c) True solutions, or molecular-ionic disperse systems, have particle dimensions in the disperse phase less than $1\ m\mu$.

P. A. Rebinder distinguishes, in addition to true solutions: a) colloidal-disperse systems with a diameter d of particles $< 0.1\mu$; b) systems with intermediate dispersion, d of the particles from 0.1 to 1μ; and c) coarsely disperse systems, d of the particles greater than 1μ.

Colloids are distinguished from true solutions by their low rate of diffusion, by their inability to pass through animal or plant membranes or parchment paper, and by their property of diffusing light. They are distinguished from coarsely disperse systems by their relative lack of stratification (settling or floating of particles).

Colloids include a) liquid colloidal solutions or sols (lyosols), which may be further divided, according to the nature of the disperse medium, into hydrous colloidal solutions, or hydrosols, and alcoholic colloidal solutions, or alcosols; b) gelatinous systems or gels, the disperse phase consisting of immobile solid particles bound to each other by molecular forces into a single mass ("massive structure") and a liquid disperse medium occupying the interstices between these particles; c) aerosols, with a gaseous (air, for example) disperse medium and a solid (smoke) or liquid (fog) disperse phase; d) solid bodies such as crystallosols (the disperse medium being a crystalline substance), certain metallic alloys, minerals formed by solidification of gels, and glass (vitreosol), in which occur innumerable isolated particles or aggregates of colloidal dimensions; single-component colloids of this kind are called isocolloids. Colloidal solutions with a solid disperse medium (crystallosols and vitreosols) are called solid dispersoids.

P. A. Rebinder [Pospelova, 1950] has distinguished the following disperse systems:

1) Disperse systems in gases

   a) with colloidal dispersion: dust in the upper layers of the atmosphere;
   b) with coarse dispersion:

   fluid in gas—fog;
   solid body in gas—smoke.

2) Disperse systems in liquids

   a) with colloidal dispersion: solid body in liquid—sol;
   b) with coarse dispersion: solid body in liquid—suspension
   gas in liquid
   liquid in liquid } emulsion

3) Disperse systems in solid bodies solid in solid—sols of gold in glass (ruby glass); colored colloidal impurities in crystals (smoky topaz, amethyst).

4) Cohesive disperse systems or massive disperse structures
   a) with a liquid interface:
   gas in liquid—foam;
   liquid in liquid—foam-like (spume-like) emulsions;

b) with a solid interface:

gas in a solid body	porous bodies—sponge, cork, activated coal;
liquid in a solid body	gels (including aerogels);
solid in a solid body	thixotropic suspensions and precipitates.

The properties of disperse systems are greatly affected by surface phenomena at the interface between the two phases.

Colloidal systems consist of aggregates made up of individual molecules and atoms (that is, they are microheterogeneous systems) or made up of huge molecules that correspond in size to colloidal particles (solutions of starch, albumin, soap, gelatin, natural and synthetic rubber, etc.). Colloids have a well-developed interface between the disperse phase and the disperse medium, and, consequently, they possess a great store of surface energy. In colloidal solutions the particles of the disperse phase are firmly bound, either directly or indirectly by an adsorbed layer of a solid substance (stabilizer), to molecules of the disperse medium, which form solvate envelopes (hydrate envelopes in hydrosols) around the particles.

The solvated (or hydrated) particles are called micelles. Colloids, to a considerable extent provisionally, are subdivided according to the degree of solvation of the particles into lyophobic colloids (hydrosols of metals, for example) and lyophilic colloids (solutions of albumin, for example).

Disperse systems in which the particles of the disperse phase, consisting of aggregates of molecules (micelles), are inactive relative to the disperse medium (i.e., mutual solubility of both phases is restricted and very small) are called lyophobic colloids, or colloids proper. In lyophobic colloids the molecular interaction between the disperse phase and the disperse medium (at the boundary between the two phases) is very weak. The particles of lyophobic colloids spontaneously coagulate, since they are not protected by solvate envelopes of the medium. The coagulation of lyophobic colloids increases with a small addition of electrolyte. It is clear that lyophobic colloids cannot be stable if a stabilizer is absent. Lyophobic colloids include hydrosols of sulfides, aqueous suspensions of soot or emulsions of paraffin in water, etc.

Hydrosols of silicic acid, hydroxides of Fe, Al, Cr, and many aluminosilicates are less lyophobic (i.e., more lyophilic); the surfaces of the particles show a greater degree of solvation of the aqueous medium, and it is for this reason that they are characterized by great aggregate stability. Thus, with a decrease in the lyophobic quality of a system, there is an increase in stability relative to coagulation, peptization increases (reverse conversion to colloidal solution—sol) in the clots, and true solutions develop.

Lyophilic colloids are characterized by a well-defined activity of the particles in the disperse phase relative to specific liquids. Such solutions involve chiefly large–molecule combinations, with molecular weights greater than 10,000 (cellulose and gelatin, for example). Consequently, lyophobic colloids are to a large or significant extent already true solutions, i.e., homogeneous systems. It is therefore clear that the practice of separating sols only into lyophobic and lyophilic is beginning to be discontinued. Most of the sols that are called lyophilic should be considered true solutions of substances with high molecular weights.

If it requires outside energy to obtain lyophobic colloids, solutions of compounds with high molecular weight are, like ordinary true solutions, formed spontaneously. Solutions of compounds with high molecular weights, like true colloidal solutions, are lyophobic (and aggregate-lyophilic) systems; they do not pass through semipermeable membranes and they diffuse slowly. Lyophobic colloids are microheterogeneous and rather unstable, whereas solutions of compounds with high molecular weights are homogeneous, stable, and they possess specific properties: tendency to swell, high viscosity, and chain molecular structure. It is proposed that substances transitional in nature, possessing properties both of molecules and of colloidal properties, be termed semicolloids.

The aggregation or coagulation of particles in colloidal solutions is produced by the addition of electrolytes, which destroy the solvate (hydrate or other) envelopes and neutralize the charge on the particles. The particles of lyophobic colloids adhere during collision and form aggregates, which, as they grow, become so much larger that they settle under the influence of gravity. Solutions of compounds with high molecular weights, consisting for the most part of lyophilic colloids, are much more resistant to the influence of electrolytes, and only when a great quantity of electrolytes is added will a flocculent precipitate form. Many lyophilic colloids, when acted on by a dehydrating agent (alcohol, acetone, etc.) change to the gelatinous state (becomes a gel) when the addition of electrolytes is large or when other factors obtain.

In a solution of colloidal particles, there is a definite charge for a given substance and solvent; the sign of the charge depends also on the pH of the environment (it is generally different in an acid environment from what it is in an alkaline environment). One may therefore speak of positive and negative colloids only in relationship to the conditions under which they generally form in the laboratory. Positive colloids (in this relative sense) are the hydrous oxides of iron, aluminum, titanium, chromium, and other metals, and basic dyes. Negative colloids generally include sulfides of copper, lead, other metals, silica, manganese dioxide, clay and humic colloids, and acid dyes.

The presence of colloidal particles all with the same sign in a solution opposes the near approach of the particles to each other, thus preventing aggregation and settling. Because of charges and great specific surface, colloidal particles of extremely small size form solvates and are found in solution in a suspended state.

In hydrophylic colloids the solid particles of the disperse phase are surrounded by stable envelopes of water molecules, which will not permit the particles to combine. During evaporation of the water, these colloids form gelatinous sediments (hydrogels) on the floor of the vessel, not dense sediments. The bond between colloidal particles and water is preserved as long as the water of the hydrogel cannot be filtered off or squeezed out; in general it cannot be removed by any means. Hydrogels include solutions of many organic substances (starch, albumin, gelatin, humic compounds), minerals of the montmorillonite group, and several others. Hydrophilic and, in general, lyophilic colloids are now considered to be true molecular solutions, but they are still studied by colloidal chemists. Many inorganic substances also form hydrosols (for example, iron oxides and SiO_2), which change to hydrogels as the water evaporates. Opal, at least in part, is a hydrogel in which particles of amorphous silica form a gelatinous mass and the disperse medium (water) occupies the interstices between particles of SiO_2.

Hydrophobic colloids are distinguished from hydrophylic by the fact that during evaporation of the water they form a dense fine-granular sediment that does not, on dilution by water, give a disperse system (i.e., does not peptize); this class of colloids includes minerals of the kaolinite group, sesquioxides, quartz, etc. Hydrophobic colloids are rather unstable, but the addition of a very small quantity of stabilizer (for example, organic acids, soap) makes them stable.

Stabilized colloids may be preserved without change for tens and hundreds of years. It is assumed that many very small crystals of solid substances (or droplets of liquid) are hydrophobic colloids, forming during crystallization or comminution (or during condensation) in an aqueous medium.

Most solutions of compounds of high molecular weights (gelatin, for example), and also several colloidal solutions (for example, hydrates of iron oxides and silicic acid), may, under certain conditions, change completely to a special solid state with no apparent separation into phases. This process is called gelatinization, and the product arising during the process is a gel. In gels the particles of the disperse phase are bound together and do not move freely as in a solution. Gels are considered to be sols having completely or partly lost their aggregate stability, but having preserved their kinetic stability. Gels include several minerals, such as agate, opal, and others. Gels that are liquid-poor or entirely dry are called xerogels. Gelatinous systems very rich in liquid are called lyogels.

Gelatinous sediments forming during coagulation of sols belong to a special group of gels: such are hydrates of iron oxides, silicic acid, and also floccules of compounds of high molecular weights during salting out of the solutions. These sediments, including a large quantity of the disperse medium, the kinetic stability of which has been destroyed, are called coagula. Last, systems similar to gels but produced from coarsely disperse suspensions are called pseudogels. Not all colloidal solutions can change into gels; the formation of gels from solutions of combinations having high molecular weight is more widespread than from certain colloidal solutions (lyophobic colloids). Elastic gels (gelatin, for example) are distinguished from inelastic gels (silica, for example). Inelastic gels absorb water that moistens them with almost no change in volume. These gels, on losing water, markedly change their physical properties and become brittle. The absorption of liquid by elastic gels is accompanied by swelling, which may be limited (with no peptization) or unlimited (with peptization).

Gels may be divided into two groups: 1) irreversible gels, which, after desiccation, do not swell when combined with the solvent a second time; and 2) reversible gels, which are able to swell even after desiccation. Examples of gels of the first group are the hydrates of metallic oxides or relatively insoluble salts. We should note that this subdivision is tentative, since the properties of gels depend on the means of preparing them and on several other factors.

Gels that change, by crystallization, to crystalline-granular aggregates are called metacolloids. In general the prefix "meta" (after) indicates a later change in a substance, and it is attached to the name of that substance. P. A. Rebinder [Pospelova, 1950] designated the following groups and subgroups of disperse systems on the basis of intensity of molecular reaction at the interface of the phases: 1) lyophilic-a) true lyophilic, b) surface lyophilic; 2) lyophobic.

In true lyophilic systems the disperse phase has completely permeated the disperse medium; i.e., the system is, as it were, a single phase (solutions of rubber in benzene, albumin and gelatin in water, etc.).

In surface lyophilic systems the particles of the disperse phase form a mosaic of lyophilic and lyophobic zones; they react with the disperse medium only from the surface. In other words, surface lyophilic systems suggest lyophobic systems in which the lyophobic particles are covered with lyophilic films, so to speak. Examples of lyophobic systems are emulsions of oil in water, hydrosols of metals, hydrosols of sulfides, and so forth.

Lyophilic (hydrophilic and other) systems are stable; they are stable in time, possibly existing for long, geologic periods. On the other hand, lyophobic (hydrophobic and other) systems are unstable, being gradually, but rather quickly, destroyed, the disperse phase separating out during coagulation of the particles under the influence of molecular forces of cohesion. It is obvious that only lyophilic colloids may preserve their properties throughout geologic time, not only during diagenesis of sediments, but, in finely dispersed clay deposits, frequently even to the stage of katagenesis (epigenesis).

As P. A. Rebinder has emphasized, free-disperse systems should be distinguished from bound-disperse systems. In free-disperse systems, or sols, the particles of the disperse phase do not form continuous rigid structures; they show no resistance to shearing. In bound-disperse systems, or gels, the particles of the disperse phase form a rigid spatial structure — a network or framework. A distinction is made between "coagels," which formed during coagulation, with stratification of the system (a lower layer of sediment and an upper layer of "cream"), and lyogels or gels, in which the homogeneity of the disperse system is preserved as a result of molecular cohesion during increasing concentration of the disperse phase. Gelatinization is a special type of coagulation, characteristic of lyophilic systems; peptization is a process in general opposed to coagulation, during which sols are formed from gels.

A scheme of the processes associated with colloidal systems (according to P. A. Rebinder) follows:

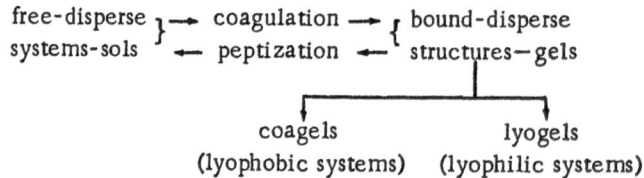

Chemical reactions in gels, because of the lack of convection currents, occur without intermixing. When insoluble compounds form, crystallization procedes slowly and quietly, and, because of this, crystals may attain great size.

"The diffusion of any substance AB from a solution into a gel containing a substance CD that is capable of reacting to produce an insoluble substance:

$$AB + CD = \frac{AC}{\downarrow} + BD$$

leads to the formation of sediment in the form of a system of layers or zones following one after the other" [Balezin and Parfenov, 1956, p. 346]. The layers of sediment are generally precipitated in increasing, rarely in decreasing, intervals of thickness.

102

" These diffusion systems that have plane surfaces give sediments in the form of a concentric system of rings. More frequently there occurs a more complex form of sediment distribution, various types of thickening, forking, symmetrically arranged strictly along radii" [Balezin and Parfenov, 1956]. The phenomenon of periodic, or rhythmic, separation of sediment is commonly called Liesegang banding (1896). The phenomenon of rhythmic segregation of any particular compound (for example, of various valent forms of iron minerals) by diffusion in silica gel is known in nature and clearly originates, not in a single plane, but in three-dimensional space, though at times from an axial line, at others from a plane, and at still others from an irregular surface. L. V. Pustovalov considers the basis of rhythmic separation of colloids to occur in a coarsely disperse medium, in a solid porous medium, or, sometimes, in the cyrstalline lattice [1932, 1940].

Chapter V

COLLOIDAL AND METACOLLOIDAL SEDIMENTARY
MINERALS AND THEIR ORIGIN

Coagulates, i.e., sediments of colloidal solutions, may settle out in a fine-granular or a gelatinous form; the latter are called gels. The coagulates of lyophobic (and hydrophobic) colloids are fine-granular, rarely flocculent precipitates. In particular, coagulates of the hydrous oxides of iron and aluminum are generally flocculent. The coagulates of lyophilic (and hydrophilic) colloids are gelatinous or very bulky flocculent sediments. The gels of hydrophilic have a very high water content. Some substances, such as silica, may form either gelatinous or fine-granular coagulates, depending on the environment. The coagulation of hydrosols, widespread in nature, leads to the formation of hydrogels. Gelatinous (the first time) coagulates of lyophilic colloids are called actual gels or coagula; these arise by gelatinization of a solution, with no separation from the solvent. Coagulators may cause gelatinization (i.e., yield coagel) in inorganic colloids such as hydrous silica, aluminum hydroxide, etc. It is quite natural that the consistency of a gel depends on the content of water in it.

The water in hydrogels of lyophilic colloids occurs as residual solvent (disperse medium) and in bound form; the amount of bound water depends on the pH value. The bound water forms envelopes (layers of oriented dipole molecules) about colloidal particles, and may even impregnate them. The more hydrophilic the colloid, the higher is the content of bound water in the hydrogel.

It should be emphasized that most coagulates have crystalline structure, as shown by x-ray and electron-microscopic studies. It has been discovered that the means by which the coagulate is obtained determines the degree of crystallization.

During chemical weathering, water participates chiefly in hydrolytic reactions, though partly as a solvent. A number of authors believe that carbon dioxide is the chief agent in weathering, although its content in the atmosphere is not large at present. In the geologic past the CO_2 content in the atmosphere was considerably greater, and the role of carbon dioxide in weathering was consequently greater.

In considering the effect of pH in a medium on the degree of dehydration of the hydrogel, let us recall that, according to Goldschmidt [1931], the following average pH values are characteristic of the principal groups of natural waters:

Waters in alkaline soils 10
Sea water 8
River water . . . . 7 (but may be even lower)
Atmospheric water 6
Swamp water . . . . 4 (generally from 2.5 to 6.0)

Note should be taken of the important role of humic colloids during weathering of minerals and rocks where a well-developed plant cover is present. Humic colloids, arising by incomplete decomposition of plant remains, are protective colloids relative to such colloidal products of chemical weathering of silicates as iron oxide and aluminum oxide, facilitating their transfer for greater distances. They also adsorb electrolytes, which are dissolved from weathered minerals and rocks.

I. I. Ginzburg [Ginzburg, 1946, and Ginzburg and Rukavishnikova, 1951] noted that silicates with sheet or chain arrangements of the silica tetrahedrons may undergo weathering in stages, with the formation of various intermediate products from the initial mineral to the final weathering products (depending on changes in pH with time). For example, muscovite changes in stages through gamma-hydromica (illite) to kaolinite, etc.

Some authors have expressed their views that the various feldspars weather differently under identical conditions: alkalic feldspars change to kaolinite, whereas plagioclase, especially that near anorthite, changes to halloysite, allophane, and montmorillonite. These views are reflected in the first edition of "Petrography of Sedimentary Rocks" by M. S. Shvetsov [1934], where the author remarks that the outer kaolin-like decomposition product of plagioclase is most likely composed of minerals of the allophane group, being gel-like and completely transient during hydrochloric acid extraction. Many authors, including F. V. Chukhrov [1955], have noted numerous examples of kaolinite forming from basic plagioclase and halloysite coming from potassium feldspars. These authors explain the formation of the various clay minerals by different conditions of weathering, both regional (climatic, etc.) and local.

It seems proper to us to consider these and other factors. In agreement with I. I. Ginzburg, most lithologists and mineralogists believe that the type of weathering and local variations in the profile of the weathering zone depend both on the composition of the rock and on the climate, and that small variations in the weathering profile are associated with local conditions of water circulation and local changes in the chemical behavior of these waters.

In speaking of the weathering zone, it should be noted that there are two principal types of such zones: areal or surface (ordinary) weathering zones and linear zones, extending to great depth (hundreds of meters) and restricted to fracture zones or to contact zones of the various rocks.

Many believe that sedimentary colloidal minerals develop chiefly in the supergene zone, the weathering zone of rocks. Despite the widespread view concerning the abundance of colloids in the weathering zone, this by no means corresponds everywhere to the facts. According to I. I. Ginzburg, the colloidal and subcolloidal fraction amounts to no more than 20% of the material in the ancient weathering zone of the Urals, the bulk of the material belonging to coarser fractions. The so-called pelicanitic crust, a weathering zone consisting of opal and kaolinite (but commonly almost exclusively opal) and forming by weathering of feldspars, is actually rich in colloids. This kind of weathering crust is developed in the Ukr SSR, the steppe of Kazakhstan, and in several other localities.

Characteristic colloidal minerals in the zone of weathering include opal, halloysite, allophane, alumina gel, several forms of psilomelane and wad, and other minerals. However, there is generally a distinct predominance of metacolloidal minerals in zones of weathering (i.e., gels that have crystallized to crystalline-granular aggregates); more properly, minerals are present that are found in metacolloidal aggregates: montmorillonite, beidellite, ferrimontmorillonite, numerous iron hydroxides of the limonite group (hydrogoethite), several minerals of the psilomelane and wad group, chalcedony, magnesite, garnierite, and others.

The oxidation zones of various ore deposits, representing distinctive and local types, exhibit weathering zones that are characterized by a whole group of colloidal and metacolloidal minerals. Colloidal minerals from such oxidized zones are [Chukhrov, 1955] allophane, asbolane, wad, vernadite, halloysite, opal, picite, and others. Metacolloidal minerals from the oxidized zone include beidellite, hydrogoethite (limonite), malachite, montmorillonite, ferrimontmorillonite, psilomelane, chalcedony, chrysocolla, and others.

Colloids of continental sedimentary origin are found in stream, lacustrine, and paludal deposits.

Silica gives a stable colloidal solution in stream waters; other authors think silica is more likely transported in nature in the form of true solution. Lithologists should consider both possibilities, as well as the possibility that it is transported under a protective coating of organic colloids (humic and other substances). Iron (in the hydroxide form) is transported by stream waters in the colloidal state (at pH values greater than 3). The transport of colloidal particles of iron hydroxides and of clay particles is facilitated by the protective action of colloidal silica or colloidal humic material. Some investigators state that silica and iron may be carried to the sea by streams that are rich in organic material. It is suggested that in these cases complex organic-mineral combinations are present and, chiefly, that the transport occurs in the colloidal form, protected by organic colloids (humic sols and, apparently, several organic acids).

Free alumina in lateritic weathering zones may have been transported by Paleozoic and Mesozoic streams both in true solutions and in colloidal solutions of clay, protected by organic material. It is also possible that alumina from ancient laterites was transported in the form of suspensions produced by mechanical comminution of the laterites. In present-day rivers, alumina is transported chiefly in a bound form, as colloidal and suspended clay particles; these are colloidal or finely dispersed "transit-born" minerals, i.e., formed during transport. Today alumina is carried partly in the free state, as colloids, only in tropical streams in lateritic regions. Under appropriate conditions, iron, being transported by river waters, may be deposited in alluvial and, in part, channel sediments, forming alluvial deposits of oolitic, oolite-like, and other iron ores. This type of iron-ore deposit has been studied in detail by L. N. Formozova [1956] in the Tertiary rocks of the Aral-Turgai lowland; these deposits occur in lenses in old erosional valleys.

Humic and other organic colloids play a protective role relative to clay colloids, making them more stable (less susceptible to the influence of coagulators). River waters, on reaching the sea, generally precipitate their colloids, as a result of the influence of electrolytes, in deltas, estuaries, lagoons, and near the shore line in general. Nevertheless, some colloidal particles reach the open sea, settling at various distances from the river mouth.

The principal types of lacustrine colloids are predominantly fresh-water deposits, such as hydrous silica (partly in diatomaceous muds), iron and manganese hydroxides, organic sapropelic material, hydrotroilite (iron hydrosulfide), native sulfur, colloidal clay minerals of the montmorillonite, beidellite, or halloysite groups, and other minerals.

There is interest in present-day bog iron ores (pisolitic, coin-structure, concretionary) and especially in fossil sedimentary lacustrine oolites, pisolites and, resembling these, bauxites that are known in the Mesozoic and Cenozoic strata of Kazakhstan, the Urals, and Siberia, and also in the Tikhvin deposit of Lower Carboniferous age. Fossil bauxite pisolites commonly contain admixtures of iron oxides; transitional facies are frequently observed in the pisolites, grading into pisolitic iron ore. Most authors consider the above-indicated bauxites to be colloidal chemical precipitates of alumina hydrates, with variable admixtures of iron oxides. The alumina hydrates were precipitated in the form of pisolites, but at times they formed a structureless gel, depending on the pH of the environment. According to the experimental data of E. V. Rozhkova and N. V. Solov'ev, [1936], the pisolitic bauxites were apparently formed at pH values less than 7, particularly at values from 2.5-3.0 to 6.5-7.0.

Colloids in paludal deposits include the colloidal (gel) part of the peat mass (humic material), the segregation of typical swamp gel (dopplerite), consisting of humic material with a variable content of calcium; accumulations of gelatinous siderite have been discovered in Belorussia, Germany, and Sweden [Bushinskii, 1946]. Apparently iron is also deposited as colloidal hydroxides in some small swamps, but normally the major part of bog hydrogoethite (limonite) is probably associated with the oxidation of lenses of bog siderite (hydrogoethite) and vivianite (hydrogoethite and picite).

Colloids in marine and lagoonal deposits are rather variable. We shall limit ourselves to a brief consideration of the separate types.

1. Clay minerals in the sea (and in lakes) are chiefly products of settling of colloidal and coarsely dispersed particles; these particles are chiefly hydromicas (derived from muscovite), montmorillonite and beidellite, ferruginous varieties of the two preceding (ferrimontmorillonite and ferribeidellite), halloysite, and kaolinite.

2. Hydrous silica may be deposited chiefly in present-day seas through the life activity of organisms (diatomaceous mud, radiolarian mud), but in earlier geologic time, when the salinity of the water was lower, the normal sedimentary precipitation of colloidal silica was apparently much more extensive. In addition, both at present and in the geologic past, hydrous silica was undoubtedly deposited and accumulated as a result of submarine volcanic eruptions; it has been deposited as a gel, but a great quantity of silica in the form of a sol may occur through the development of organisms with siliceous shells. Many concretions of silica are apparently due to settling of colloidal silica, whereas others owe their origin, in part, to the presence of remains of siliceous organisms and, in part, to the precipitation of SiO_2 as a gel [Teodorovich, 1935[1], 1950].

3. Colloidal iron minerals show limited development in modern seas, but in more ancient times they formed extensively, iron colloids migrating to the open sea, probably protected by humic material. In Precambrian fresh or slightly salty seas [Strakhov, 1947], colloidal particles of iron hydroxide might have been carried far out to sea, in view of the low electrolyte content in the water, out to the pelagic zones; at that time there occurred rhythmic precipitation of colloidal iron oxides and silica, leading to characteristic finely banded

sediments (ferruginous quartzites and ferruginous hornfels of Precambrian age, which have obtained their present properties by later metamorphism).

N. M. Strakhov [1947] has noted that at the beginning of post-Algonkian time colloidal particles of iron oxides (hydroxides), because of increasing salinity of the marine waters, began to settle nearer and nearer the shore.

4. The origin of manganese colloidal minerals in the sea is sometimes explained by the life activity of microorganisms [V. I. Vernadskii, 1937]. In the USSR large deposits of sedimentary manganese ore are associated with Paleogene marine basisn; they occupy entire provinces, including the Nikopol' deposit in the Ukraine, the Chiatura and Adzhameti-Chkhari deposits in the Georgian SSR, the Polunochnoe and Marsyaty deposits in the Urals, the Mangyshlak deposit in Kazakhstan, and so forth. The manganese deposits of the USSR have been studied by A. G. Betekhtin [1937, 1944[1,2]], the Chiatura deposit in special detail, the Polunochnoe and others in some-what less detail. At Chiatura a wide gulf of the Oligocene sea, apparently having low salinity, permitted pre-cipitation of manganese in gel form and the formation of oolite-like nodules. It may be that manganese sols developed because of specific microorganisms. Hydrous silica, in the form of opal, also coagulated with the manganese oxides in the Chiatura gulf.

The waters in the Chiatura basin also contained large quantities of iron oxides, but the manganese ap-parently precipitated separately, seemingly being controlled by different values of pH for the precipitation of iron and manganese oxides [Betekhtin, 1937]. The oxide ores of manganese are primary; i.e., they were deposited in an acidic environment, whereas farther from the shore of the basin such ores give way in facies change to carbonate manganese ores.

5. Marine (littoral-marine) aluminum ores ("marine" bauxites) form by settling of colloidal alumina in freshened lagoons and marine gulfs, generally in a geosynclinal zone, forming in limestones; this type includes the Paleozoic bauxites of the Urals (Devonian) and other regions of the USSR, China (Lower Carboniferous), and several other countries. The Dvonian bauxites of the Urals [Nalivkin, 1942] have pisolitic structure, are com-posed of diaspore and boehmite, and contain admixtures of iron minerals such as hematite and hydrohematite or leptochlorite. The development of boehmite with fine-aggregate structure in the Paleozoic geosynclinal bauxites allows us to apply views concerning the origin of boehmite in platform bauxites to explain the origin of these geosynclinal boehmite occurrences (spontaneous dehydration of water-rich primary gel).

6. Sedimentary phosphorites are apparently sediments of marine waters. A. V. Kazakov [1937[2], 1939] distinguished platform (nodular) and geosynclinal (bedded) phosphorites. He pointed out that the separation of calcium phosphates and carbonates from supersaturated solutions of sea water may occur by coagulation, with finely dispersed $CaCO_3$ in such cases generally forming no more than 10% of the rock. G. I. Bushinskii [1952] believes that colloidal-amorphous (and crystalline) kurskite is the principal mineral in the nodular phosphorites; but he has found amorphous collophane (and crystallized francolite) in the bedded deposits.

7. Glauconite, as noted by several authors, also passes through a primary-colloidal state, which may form a chemical marine sediment. Glauconite grains of the normal type, during weathering or in fresh varieties, commonly display their primary structure of small globular colloidal (apparently) particles. The colloidal nature of glauconite is also indicated by the fractures of desiccated gel, observed in normal-type (lobate) grains.

L. Cayeux [1916] distinguished as many as eight types of glauconite grains in transmitted light in thin sections, only three types being found frequently: a) homogeneous glauconite, consisting of ordinary simple grains, which in transmitted light in thin sections show no break down into smaller divisions or globules and which show microaggregate extinction under crossed nicols; b) globular glauconite, the grains consisting of small spheres (globules), each of which appears as a microaggregate under crossed nicols; and c) granular glauconite, showing in transmitted light indications or clear outlines of breaking down into finer grains, and showing aggre-gate extinction under crossed nicols. In addition, transitional forms between globular and granular glauconite have been observed.

A microscopic-petrographic study of the Maestrichtian phosphorites of the Blyavinskii district, made by the author in 1935 (the deposit consisting of a nodular phosphorite layer), revealed extensive occurrence of strictly glauconitic rocks with phosphatic cement. It is interesting that the glauconitic grains commonly show a globular structure, generally emphasized by irregular coloration and apparently associated with weathering. The glauconite of the Blyavinskii phosphorites thus are principally microconcretions of very small globular, primary, colloidal (probably) bodies. In the principal, or nodular, phosphorite layer there is a clear dominance of phosphorite of the glauconitic type, consisting essentially of glauconitic rock with phosphatic cement. The glauconite grains in this layer are irregularly rounded, occasionally lobate, chiefly from 0.1 to 0.4 mm in diameter, and showing microaggregate polarization. There are two principal types of these grains: a) light green grains and b) grains with spotted coloration (light and dirty green), more or less pure green color being rarely observed.

It is possible to see in a number of the spotty grains that the light green zones consist of accumulations of very small (about 0.005 to 0.01 mm) more or less rounded bodies; i.e., they show globular structure. Each rounded body shows itself to be an aggregate or a very small indivisible unit under crossed nicols. In other parts of the spotty grains, very small light green globular bodies occur in a dirty green "field." On examining a number of light green glauconite grains, one may note that they represent a compact accumulation of very small light green globules; i.e., they possess an indistinctly defined (fused) globular structure.

The glauconite studied by us from the phosphorites of the Blyavinskii district represents a family of types, of which two are the most extensive: a) light green and green glauconite grains with microaggregate polarization, generally appearing more or less homogeneous with but one nicol in the optical system; b) spotty (light and dirty green) glauconite grains with more or less well-defined globular structure, also showing microaggregate polarization (the globular glauconite of Cayeux). The presence of light green grains with almost imperceptible globular structure forming a connecting form between these two types of glauconite also emphasizes the view of Cayeux that ordinary "granular" glauconite forms by intimate "fusion" of individual "spheres" of glauconite of the globular type.

Apparently the bulk of the Blyavinskii glauconite that shows microaggregate polarization consisted, at the culmination of its development, of intimately fused (definite outlines therefore having been lost) globular bodies of colloidal origin, the grains having had primarily a more or less pure green color.

Glauconite characterizes a definite geochemical facies [Pustovalov, 1933; Teodorovich, 1947] of oxidation-reduction conditions. As noted by F. V. Chukhrov [1955, p. 115]: "...the distinctive morphology of glauconite is in full accord with the view of metacolloidal origin."

8. Sedimentary iron chlorites, or leptochlorites, are frequently found in the matrix (we shall not now speak of oolites and oolite-like bodies of leptochlorite) of primary colloidal structure. Teodorovich [1939] demonstrated the sedimentary origin of the Khalilovotype Novo-Troitsk deposit, and he showed that the abundant iron chlorites in the ore sequence at this deposit are chemical precipitates and were primarily amorphous, i.e., colloidal in origin. This conclusion is based on the following data: a) all transitions are observed from amorphous and cryptocrystalline leptochlorite to definitely crystalline varieties (showing the crystallization of a primary gel of iron chlorite); b) the texture of the leptochlorite (the distribution of the scaly individuals) indicates that the material formed in situ; c) in the leptochlorite mass are found variously formed zones with concentrically layered structure, indicating a chemical metacolloidal origin for the leptochlorite.

9. Hydrotroilite, a black gelatinous hydrate of iron sulfide, has been found in abundance in the muds of the Black Sea and of the bordering estuaries. Hydrotroilite aggregates are generally very small and occur chiefly in the form of floccules, being found in calcareous-argillaceous or arenaceous-argillaceous sediments.

10. Such minerals as colloidal calcite (bütschliite) and colloidal dolomite (gurhofian), being found in accumulations in nature and consisting of very small pelitomorphic grains (d < 0.005 mm), generally of irregular form, are apparently rapidly crystallized sediments of lyophobic colloids. They may develop during deposition of primary pelitomorphic limestones and dolomites, but they also form during epigenesis on a small scale. An example of epigenetic development is the incrustation of cryptocrystalline $CaCO_3$ found by the author in 1936 on nodules and blocks of separate phosphorite flags in the Bogdanovka district (Aktyubinsk region, northwestern Kazakhstan), occurring in phosphorite beds immediately beneath the soil.

From studies of the Kashira primary-chemical dolomites of the upper Volga region, L. V. Pustovalov, in his manual [1940], noted the metacolloidal nature of dolomitic rocks in general. This conclusion is valid for the deposits actually studied by L. V. Pustovalov and, possibly, also for most primary pelitomorphic dolomites.

Colloidal sedimentary minerals of diagenetic (scarce) and epigenetic (more abundant) origin are distinguished from syngenetic minerals if we exclude from the latter those minerals forming during sedimentation and at the incipient stage of diagenesis. Diagenesis, i.e., the conversion of initial sediments into rock, involves the processes of compaction, dehydration, cementation, the formation of concretions, and reduction (or oxidation in the upper zones of muds, etc.). Minerals, colloidal and metacolloidal, that may form diagenetic and epigenetic concretions and cement in rocks, are: a) hydrogoethite (limonite), Fe and Mn oxides (lacustrine and marine), and also manganese oxides (of the wad group); b) opal or opal and chalcedony, changing to quartz with time; c) calcium phosphates; d) glauconite in the form of normal lobate grains, representing microconcretions; e) various carbonates (siderite, ankerite, and others), forming microconcretions or microaggregates; f) iron sulfides (hydrotroilite and melnikovite), later changing to pyrite or marcasite; g) leptochlorites, oxides of iron and alumina, forming pisolites (i.e., microconcretionary in nature).

Most of the indicated colloidal and metacolloidal minerals, especially types "a," "b," "e," and "f," may be of diagenetic as well as epigenetic origin.

Minerals that are colloids in diagenetic cementation are silica (opal), several colloids and metacolloids of the clay minerals, phosphates, and iron hydroxides.

It is of considerable interest that N. Ya. Denisov and P. A. Rebinder [1946] explain the plasticity of clay by the presence of hydrated colloidal films of silica on the particles. The dehydration of the colloidal films of silica, occurring during lithogenesis and epigenesis, makes the clays less plastic, i.e., makes them rigid. On the other hand, these authors conclude that a surface sol forms on the clay particles during reaction between the particles and water, due to peptization of clay colloid; this sol may be changed into a gel. It is the gel-producing colloidal films that cover the clay particles which bind (glue) the particles together. If such colloidal solutions are stabilized, the enveloping colloidal films acquire a lubricating quality. All these processes may be effective during epigenesis. The loss of plasticity and the solidification of clay rocks are associated by N. Ya. Denisov and P. A. Rebinder with the dehydration of the colloidal films. The variable properties of the enveloping colloidal films and, especially, the variable tendency of these films to dehydrate apparently explain why certain old clays have preserved their plasticity (some in the Leningrad district, for example, and in Esthonia); on the other hand, this variability may be the cause of the lack of plasticity in certain clays that are geologically young (the Maikopian clays of the northern Caucasus, for example). It seems to us that, in this latter and similar cases, the loss of plasticity may be related to two factors: the strength of tectonic pressure and the very great total weight of overlying deposits (not only preserving them till the present day but preventing erosion).

The plasticity of clays at present is generally associated with envelopes of hydrate ions around the particles. The plasticity of clay rocks or sediments is higher the thicker these films are and the greater the specific surface area of the rocks, i.e., the smaller the particles forming them. In other words, the plasticity of clay deposits increases with increase in the clay fraction, i.e., with increase in finer, especially colloidal, particles. The plasticity of clay rocks generally depends on the properties of the solid particles, the mineral composition, their size and form, the kind of exchange cations, and also the reaction between the particles and the liquid (generally aqueous) phase, the mineralization of this phase, its pH, and its oxidation-reduction character.

We should also note here the development of colloidal films ("jackets") of oxides and hydroxides of Fe or Fe and Mn on the surface of sand grains, which form in sandy rocks during epigenesis (i.e., in lithified rocks) and sometimes during diagenesis. These films consist chiefly of iron oxides and hydroxides (or in combination with manganese oxides and hydroxides, occasionally with manganese compounds as the dominant constituent), and partly of alumina and silica. Normally the formation of such iron or iron-manganese films is associated with the segregation of corresponding gels from colloidal or true solutions in circulating ground waters within lithified rocks or from initial soil or subsoil waters. Commonly the formation of sandstones with cement of iron or iron-manganese oxides may be associated with fluctuations of the water table, being observed within the interval between the probable highest and lowest position of this level.

Let us pause very briefly on the question of dendrites, among which crystalline and coagulated forms are distinguished. "Dendrites" is a term applied to flat tree-like formations or growth figures sometimes observed on the surfaces of rocks or minerals, along fractures or cleavage. Dendrites are commonly flat dendritic crystals, forming by rapid crystallization; the development of these has been studied in steel.

Two types of coagulated dendrites, associated with colloidal solutions, are distinguished: the most common is two-dimensional, or flat, the other three-dimensional. Two-dimensional dendrites form in joints and on bedding planes, and generally consist of iron or manganese hydrous oxides or of a combination of the two; they are characteristically black, brownish, or rust-colored. Psilomelane dendrites are especially typical; they are sometimes mistaken for plant imprints.

The development of two-dimensional dendrites, observed in the fractures of certain rocks, is generally explained by the peculiarities of spreading of colloidal solutions along the surfaces of joints or fractures. It is well known that a thin layer of gel squeezed between two glass plates (until they adhere) will give dendritic figures when the plates are separated quickly. The principal cause for coagulation of colloids in dendrites is generally assumed to be the evaporation of water or the absorption of water through capillary openings in rocks and minerals.

Three-dimensional coagulated dendrites form during transfer and rhythmic deposition of one colloidal substance in the mass of another. Sometimes three-dimensional dendritic figures develop in gels because of the growth of membranous tubular cells; the origin of dendrites in some agates may be explained in this way (see p. 103 concerning Liesegang rings).

•   •   •

We have already noted the possibility of certain colloidal sediments, under certain conditions, becoming liquid, changing from a gelatinous state to a sol or to a suspension (peptization) when shaken or stirred. After the action ceases, the colloidal or suspension system becomes quiet, and the precipitation of a gel is again observed.

All these phenomena, called thixotropic in colloidal chemistry, are very important in the formation of clay sediments and the formation of particular clay minerals. The time of the reverse change from the liquid state to a plastic state is definite for a given system, and is a measure of its thixotropic properties. The structure-producing and the general structural-mechanical, or thixotropic, properties of clay have been studied by many authors. N. N. Serb-Serbina and P. A. Rebinder [1947], for example, discovered three types of shearing stress (deformation) curves for suspensions of bentonitic clays: a) representing structures with coarse fractures (faulted structures); b) representing structures with plastic disturbance; and c) curves of plastic flow.

Coagulation and peptization are mutually reversible processes in colloidal systems and suspensions. Coagulation in colloids and suspensions is effected by consolidation of the particles, and this leads to accelerated settling, i.e., to more rapid sedimentation. Hydrophobic colloids (minerals in the kaolinite group, sesquioxides, and others) coagulate much more easily and quickly than hydrophilic colloids (minerals in the montmorillonite groups, humic substances, etc.). One should not forget that it is more proper to speak of colloidal and crystalloidal states of a substance, and these are determined by environmental conditions. Consequently, in referring a given substance to colloids in general and to hydrophilic colloids in particular (or to crystalloids, etc.) we are designating merely the state of the substance, the state in which it is generally observed in nature or in the laboratory.

SUMMARY

This consideration of the principal and less abundant authigenic minerals in sedimentary rocks (except for rare forms) makes it possible to draw some conclusions concerning the great variety among the typical minerals of sedimentary rocks and also concerning the development of these minerals at different phases and stages of formation of sedimentary deposits.

It seems to us that the description of the authigenic minerals in sedimentary rocks according to the plan we have adopted is expedient, permitting the reader to orient himself easily. The outline, in its most general form, is as follows:

1. Silica group.

2. Carbonate group (anhydrous, basic, and hydrous).

3. Sedimentary silicate group.

   Iron silicates:

      Potassium-bearing (glauconite group).

      Non-potassium (iron and iron-magnesium chlorites).

   "Clay group" silicates.

   Calcium and sodium hydrous aluminosilicates (zeolites).

   Authigenic feldspars.

4. Oxide and hydroxide group (Ti, Al, Mg, Fe, Mn, and Cu).

5. Sulfide group (Fe, Cu, Mn, Pb, and Zn).

6. Phosphate group.

   Calcium phosphates.

   Iron phosphates.

7. Sulfate and fluoride group (fresh-water, brackish, and marine).

8. Soluble salt group (lacustrine and lagoonal salts).

9. Native element group.

In Chapter III we described the authigenic minerals of sedimentary rocks, including in the descriptions the chemical composition and characteristic properties of each mineral, the properties distinguishing one mineral from similar minerals, the origin, and the distribution in rocks and mineral deposits. Data have been given on the physicochemical conditions under which the minerals form, and for certain minerals note has been made of the normal restriction of syngenetic forms of the minerals to definite geochemical facies and of epigenetic forms to corresponding secondary physicochemical conditions; in some cases the optimum values of salinity and temperature have been pointed out. A review of the existing data indicates that most minerals in sedimentary rocks may form during syngenesis (more precisely, during diagenesis of sediments) as well as during epigenesis, but still there are a great number that are typical minerals of diagenesis of sediments. Some of the authigenic minerals in sedimentary rocks are chemical precipitates from waters in the parent basin, the grains of which grew and became cemented during diagenesis. There are a number of authigenic minerals associated predominantly with surface weathering or oxidation of formational waters during epigenesis. On the other hand, individual epigenetic minerals commonly owe their development to local reducing processes (the introduction of carbon dioxide gas, of petroleum, of hydrogen sulfide formational waters, etc.).

The development of sedimentary rocks is multiform: they pass through various stages of growth and in different orders (see Table 1, p. 12). The history of a sedimentary rock and of the processes of authigenic mineral growth is made even more complex when the deposits are subjected to incipient metamorphism (when metamorphism is distinct or marked, metamorphic rocks develop).

The fundamental task for mineralogists and lithologists involves reconstructing the geologic history of each rock studied by discovering the sequence and the time of formation of all the principal or characteristic minerals. In doing this it is entirely unavoidable that paragenetic groups of minerals will be distinguished (early, middle, and late diagenesis of sediments; early and late katagenesis, hypergenesis, incipient metamorphism, etc.).

LITERATURE CITED

E. A. Abramova, Author's abstract of dissertation: Processes of Authigenic Mineral Formation in Sandy Rocks of the Devonian System in the Upper Volga Region (Kuibyshev and Saratov Districts) [in Russian] (Petroleum Institute, Acad. Sci. USSR, 1954).

T. P. Afanas'ev, "The role of ground water in the dolomitization of rocks," Doklady Akad. Nauk SSSR, 62, 4 (1948).

P. F. Andrushchenko, "Mineralogy of the manganese ores of the Polunochnoe deposit. Tr. Inst. geol. Nauk, AN, AN, SSSR, 150, seriya rudn. mestorozhd., 16 (1954).

O. M. Ansheles and V. B. Tatarskii, "Enlargement of feldspars in Devonian sands," Izv. Glavn. geol.-razved. upr., 50, 25 (1931).

A. D. Arkhangel'skii, "Iron sulfide in Black Sea sediments," Byull. Mosk. obshch. ispyt. prirody, otd. geol., 12, 3 (1934).

P. P. Avdusin, "Petrographic correlatives of the Cretaceous and Jurassic rocks in the southwestern part of the Ural-Emba district," Neftyanoe khozyaistvo, 2 (1938).

P. P. Avdusin, Mud Volcanoes in the Crimea-Caucasus Geologic Province (Petrographic Investigations) [in Russian] (Acad. Sci. USSR Press, 1948).

S. A. Balezin and G. S. Parfenov, Fundamentals of Physical and Colloidal Chemistry [in Russian] (Moscow, Gosuchpedizdat, 1956).

F. A. Bannister and W. F. Whittard, A magnesian chamosite from the Wenlock limestone of Wickwark, Glouchester-shire. Min. Mag., 1945, 27.

V. P. Baturin, "Albitization of certain sedimentary rocks in the region of the Georgian Military Highway," Izv. Geol. kom., 47, 1 (1928).

V. P. Baturin, Reference Manual on the Petrography of Sedimentary Rocks, [in Russian] (Moscow-Leningrad, ONTI, 1932), pt. 1.

V. P. Baturin, "The stability and synthesis of minerals in the deep geosphere and sedimentary shell," Doklady Akad. Nauk SSSR, 37, 1 (1942).

V. T. Belousova, Determining the Mineral Content of Clay Rocks by Immersion and Staining Methods [in Russian] (Goztoptekhizdat, 1948).

D. S. Belyankin, "A description of the mineral monothermite," Doklady Akad. Nauk SSSR, 18, 9 (1938).

D. S. Belyankin, V. V. Lapin, and I. A. Ostrovskii, "Investigations on dolomitized limestones in polished sections in reflected light," Izvest. Akad. Nauk SSSR, seriya geol., 2 (1940).

D. S. Belyankin and V. P. Petrov, "The petrography and petrogenesis of the Askani clays," Izvest. Akad. Nauk SSSR, seriya geol., 2 (1950).

L. G. Berg, A. V. Nikolaev, and A. Ya. Rode, Heating the Cooling Curves [in Russian] (Moscow-Leningrad, Acad. Sci. USSR Press, 1944).

A. G. Betekhtin, The Origin of the Chiatura Manganese Deposit- Transactions of the Conference on Origin of Iron, Manganese, and Aluminum Ores [in Russian] (Acad. Sci. USSR Press, 1937).

A. G. Betekhtin (1), "The genetic types of manganese deposits," Izvest. Akad. Nauk SSSR, seriya geol., 4 (1944).

A. G. Betekhtin (2), Industrial Manganese Ores in the USSR [in Russian] (Acad. Sci. USSR Press, 1944).

A. G. Betekhtin, Mineralogy [in Russian] (Moscow, Gosgeolizdat, 1950, also 1956).

Z. A. Bogdanova, The Significance of Various Lithologic Factors in the Qualitative Description of Limestone and Dolomite Deposits [in Russian] Symposium of the VSEGEI on Lithology in Memory of Prof. S.F. Malyavkin, 1 (1940).

A. M. Boldyreva, "Authigenic analcime from the Upper Permian deposits of the Chkalov and Aktyubinsk districts," Zap. Min. obshch., 82, 4 (1953).

P. A. Borisov, "Crystals of feldspars and mica in dolomites from the environs of Povenets," SPb. Obshch. estestvoizpyt., 11, 1 (1909).

R. Brauns, Chemical Mineralogy [in Russian] (St. Petersburg, Rikker Press, 1909).

H. T. Britton, Hydrogen Irons [Russian translation from English] (ONTI, 1936).

S. V. Bruevich, The oxidation-reduction potential and pH of sediments in the Barents and Kara seas," Doklady Akad. Nauk SSSR, 19, 8 (1938).

K. Buch, Die Kohlensäurefaktoren des Meerwassers. Rapports et proces. verbaux des reunions, vol. 67. Conseil permanent international pour l'exploration de la mer. Copenhague, 1930.

E. Z. Bur'yanova, "Analcime-bearing sedimentary rocks in Tuva," Doklady Akad. Nauk SSSR, 98, 2 (1954).

E. Z. Bur'yanova, Authigenic Laumontite from the Middle Devonian Sandstones of Tuva [in Russian] Information Circular of the VSEGEI, 3 (Gosgeoltekhizdat, 1956).

G. I. Bushinskii, "Conditions of accumulation of siderite, vivianite, and brown iron ores in the swamps of Belorussia," Byull. Mosk. obshch. ispyt. prirody, otd. geol., 21, 3 (1946).

G. I. Bushinskii, "Mordenite in Jurassic, Cretaceous, and Paleogene marine deposits," Doklady Akad. Nauk SSSR, 73, 6 (1950).

G. I. Bushinskii, Apatite, Phosphorite, and Vivianite [in Russian] (Acad. Sci. USSR Press, 1952).

G. I. Bushinskii, "Lithology of the Cretaceous deposits in the Dnepr-Donets basin," Tr. Inst. geol. nauk AN, SSSR, 156, geol. seriya, 67 (1954).

G. I. Bushinskii, "Calcium phosphates in phosphorites," in Problems on the Geology of Agricultural Ores [in Russian] (Moscow, Acad. Sci. USSR Press, 1956).

G. I. Bushinskii and V. A. Frank-Kamenetskii, "The hydraulic activity and x-ray characteristics of opaline material in tripoli and diatomite," Doklady Akad. Nauk SSSR, 96, 4 (1954).

L. Cayeux, Introduction à l'ètude pétrographique des roches sédimentaires. Paris, 1916.

V. N. Chirvinskii, "Podolite, a new mineral," Centralblatt Min. (1907).

V. N. Chirvinskii, "Feldspatization of the Kiev chalk," Geol. vestnik, 2, 3 (1916).

V. N. Chirvinskii, Phosphorites of the Ukraine [in Russian] (Kiev, 1919).

F. V. Chukhrov, Colloids in the Earth's Crust [in Russian] (Moscow, Acad. Sci. USSR Press, 1955).

V. F. Chukhrov and F. Ya. Anosov, "The nature of chrysocolla," Zap. Vses. min. obshch., 79, 2 (1950).

F. V. Chukhrov and L. P. Ermilova, "New data on kerchenites," in the Collection: Questions on Geochemistry and Mineralogy [in Russian] (Acad. Sci. USSR Press, 1956).

E. S. Dana, Descriptive Mineralogy (Manual) [Russian translation from English, under the editorship and with the assistance of A. E. Fersman and O. M. Shubinkova] (ONTI, 1937).

J. D. Dana, E. S. Dana, C. Palache, H. Berman, and C. Frondel, System of Mineralogy [Russian translation from English] (Moscow, IL, 1, pts. 1 and 2, 1951; 2, pt. 1, 1953; 2, pt 2, 1954).

N. Ya. Denisov and P. A. Rebinder, "The colloidal-chemical nature of the bond in clay rocks," Doklady Akad. Nauk SSSR, 50, 6 (1946).

M. G. Dyadchenko and A. Ya. Khatuntseva, "New data on leucoxene," Geol. zhurnal, 14, 4 (1954).

M. G. Dyadchenko and A. Ya. Khatuntseva, "The formation of glauconite in a continental environment," Zap. Vses. min. obshch., seriya 2, 85, 1 (1956).

G. S. Dzotsenidze, "Analcime of sedimentary origin from the Bathonian carbonaceous shales in the vicinity of Kutaisi," Soobshch. Akad. Nauk SSSR, 4, 10 (1943).

K. O. Emery and S. C. Rittenberg, Early diagenesis of California basin sediments in relation to origin of oil.- Bull. Amer. Assoc. Petrol. Geol., 1952 36, N. 5.

E. P. Ermolova, Author's abstract of dissertation: Secondary Mineralizing Processes in the Oligocene and Miocene Sandy Rocks of Georgia (Petroleum Institute, Acad. Sci. USSR, 1952).

E. P. Ermolova, "The sequence of the mineral-forming processes in the Miocene and Oligocene sandy rocks of Georgia," Doklady Akad. Nauk SSSR, 90, 2 (1953).

E. P. Ermolova, "The formation of authigenic minerals in the Miocene and Oligocene sandy and silty rocks of Georgia" in the Collection: Data on the Geology and Petroleum Potential of Georgia [in Russian] (Acad. Sci. USSR Press, 1956).

K. Fadeev (K. Thaddeeff), "Remarks on some reactions for identifying minerals," Zeitschrift Krist., 20 (1892), pp. 348-353.

M. I. Fadeev, "A technique for determining magnesite in solid and unconsolidated samples," Za Bashkirskuyu neft', 6 (1936).

M. I. Fadeev, "The identification of crystalline magnesite under the microscope," Zap. Vses. min. obshch., seriya 2, 84, 3 (1955).

F. Feigl u. Leitmeier. Eine Reaktion zur Unterscheidung von Dolomit und Magnesit.–Cbl. Min., Geol. u. Palaontol., Abt. A., 1928.

A. E. Fersman, The Geochemistry of Russia, 1 [in Russian] (Petersburg, Scientific Chemical-Technical Press, 1922).

A. E. Fersman, Geochemistry, vol. 2 [in Russian] (Leningrad, ONTI, Khimteoretizdat, 1934).

A. E. Fersman, Geochemistry, vol. 3 [in Russian] (Leningrad, ONTI, Goskhimtekhizdat, 1937).

A. E. Fersman, Geochemical and Mineralogical Methods of Prospecting for Mineral Deposits [in Russian] (Moscow-Leningrad, Acad. Sci. USSR Press, 1939).

E. E. Flint, Beginning Crystallography [in Russian] (Gosgeolizdat, 1952).

L. N. Formozova, "Glauconitic sands in the Kyzyl-Sai district," Tr. Inst. geol. nauk AN SSSR, 112, geol. seriya, 38 (1949).

L. N. Formozova, "Composition and conditions of formations of the oolitic iron ores in the deltaic facies of the Middle Oligocene deposits of the Aral region," Izvest. Akad. Nauk SSSR, seriya geol., 5 (1953).

L. N. Formozova, Author's abstract of dissertation: Iron Ores in the Northern Aral Region [in Russian] (Geological Institute, Acad. Sci. USSR; 1956).

E. K. Frolova, "Magnesite in the Lower Permian rocks of the Volga region near Kuibyshev and Saratov," Izvest. Akad. Nauk SSSR, seriya geol., 5 (1955).

E. P. Furman, "The mineralogy of the phosphorite deposits in the Dnestr region," in the Collection: Questions on the Mineralogy of Sedimentary Rocks, Book 1 [in Russian] (Lvov University Press, 1954).

Ch. M. Gilbert and M. G. McAndrews. Authigenic heulandite in sandstone Santa-Crus country, California.– Journ. Sediment. Petrol. 1948, 18, N 3.

I. I. Ginzburg, "Geochemistry of the weathering zone on serpentinites in the southern Urals," Izvest. Akad. Nauk SSSR, seriya geol., 1 (1938).

I. I. Ginzburg, "Types of montmorillonitic and halloysitic weathering of rocks," Tr. Inst. geol. nauk AN, SSSR, 41, seriya rudn. mestorozhd., 5 (1941).

I. I. Ginzburg, "Some physicochemical events in the formation of bauxite deposits," Izvest. Akad. Nauk SSSR, seriya geol., 4 (1942).

I. I. Ginzburg, The geochemistry and geology of ancient weathering zones in the Urals," Tr. Inst. geol. nauk AN SSSR, seriya Ural'sk eksp., No. 2, 81 (1947).

I. I. Ginzburg and I. A. Rukavishnikova, Minerals in Ancient Weathering Zones in the Urals [in Russian] (Moscow, Acad. Sci. USSR Press, 1951).

V. M. Goldschmidt, Der Kreislauf der Metalle in der Natur.– Metallwirtsch., Metallwiss., Metalltechn., 1931, 10, N 14.

N. I. Gorbunov, I. G. Tsyurupa, and E. A. Shurygina, X-ray Photographs, Thermal Diagrams, and Dehydration Curves for Minerals Found in Soils and Clays [in Russian] (Acad. Sci. USSR Press, 1952).

L. I. Gorbunova, "Glauconites in the Jurassic and Lower Cretaceous deposits in the central part of the Russian platform" Tr. Inst. geol. nauk AN SSSR, 114, geol. seriya, 40 (1950).

R. E. Grim, Clay Mineralogy [Russian translation from English] (Moscow, IL, 1956).

G. V. Gvakhariya, Zeolites of Georgia [in Russian] (Tbilisi, Acad. Sci. Georgian SSR Press, 1951).

A. Hadding, The pre-Quaternary sedimentary rocks of Sweden. IV. Glauconite and glauconitic rocks. Lund 1932.

W. F. Hillebrand and G. E. Lendel, A Practical Guidebook to Inorganic Analysis [Russian translation from English] (Moscow, ONTI, 1935).

V. P. Ivanova, "Chlorites," Tr. Inst. geol. nauk AN SSSR, 120, petrograf. seriya, 38 (1949).

M. I. Kantor, Origin of the Kerch Ores [in Russian] (K. A. Timiryazev Agricultural Academy Press, 1938).

L. I. Karyakin and N. V. Logvinenko, "Concerning the papers of Dyadchenko and Khatuntseva on continental glauconite," Zap. Vses. min. obshch., 86, 3 (1956).

A. V. Kazakov (1), "The chemistry of the natural phosphatic substance in phosphorites and its origin: 1. The system $CaO-P_2O_5-H_2O$ in the regions of low concentrations," Tr. Nauchn. inst. udobr. i insektofungisidov (NIUIF), 139 (1937).

A. V. Kazakov (2), "The phosphorite facies and the origin of phosphorites," Tr. NIUIF, 142 (1937).

A. V. Kazakov, "The lithology and weathering processes in phosphatic cores from the Egor'evsk group of deposits: Phosphates of the Moscow district," Tr. NIUIF, 140 (1938).

A. V. Kazakov, "The phosphate facies: 1. The origin of phosphorites and the geologic factors involved in their accumulation," Tr. NIUIF, 145 (1939).

A. V. Kazakov and L. I. Gorbunova, Glauconite as an Indicator of Facies [in Russian] (Abstracts of scientific-research work for 1945, Division of the geologic-geographical sciences, Acad. Sci. USSR Press, 1947).

A. V. Kazakov, M. M. Tikhomirova, and V. I. Plotnikova, "The system of carbonate equilibria (dolomite, magnesite)," Tr. Inst. geol. nauk AN SSSR, 152, geol. seriya, 64 (1957).

N. M. Knipovich, Hydrology of the Seas and Saline Waters [in Russian] (Pishchepromizdat, 1938).

M. M. Konstantinov, "Exogenetic lead and zinc sulfides," in the Collection: Questions on the Mineralogy of Sedimentary Rocks, Book 1 [in Russian] Lvov University Press, 1954).

A. G. Kossovskaya, "Lithologic-mineralogical characteristics and the conditions of formation of clay in the Produktivnyi sequence of Azerbaidzhan," Tr. Geol. inst. AN SSSR, 153 (1954).

A. G. Kossovskaya and V. D. Shutov, "Epigenetic zones in the clastic section of Mesozoic and Upper Paleozoic rocks in western Verkhoyan'e," Doklady Akad. Nauk SSSR, 193, 6 (1955).

A. G. Kossovskaya and V. D. Shutov, "The character and distribution of authigenic minerals in the Mesozoic-Paleozoic section in western Verkhoyan'e," Tr. Geol. inst. AN SSSR, 5 (1956).

B. P. Krotov, "Dolomites, their formation, conditions of stability in the earth's crust, and alteration, as related to a study of the dolomites in the upper layers of the Kazan series in the vicinity of Kazan," Tr. Obshch. estestvoispyt. pri Kazansk. univ., 50, 6 (1925).

B. P. Krotov, "Iron chlorites of the strigovite type," the the Collection: Iron-Ore Deposits of the Alapaevsk Type on the Eastern Slope of the Urals, and Their Origin 2 [in Russian] (Moscow-Leningrad, Acad. Sci. USSR Press, 1936).

B. P. Krotov, "Discovery, in the Khalilovo iron-ore deposit, of magnetite having formed from solutions of surface origin at low temperatures," Doklady Akad. Nauk SSSR, novaya seriya, 26, 8 (1940).

B. P. Krotov, "Systematic relations in the deposition and distribution of the near-shore parts of marine manganese, iron, and aluminum ores," Zap. Min. obshch., 72, 1 (1943).

M. I. Kuadzhe, The Nal'chik Bleaching Clays [in Russian] (Moscow-Leningrad, The People's Committee on Local Industry RSFSR Press, 1938).

M. A. Kulakov, "Colloidal substances in natural waters," Izvest. Akad. Nauk SSSR, Otd. mat. i estestv. nauk, 1 (1938).

N. S. Kurnakov and V. I. Nikolaev, "Solar evaporation of marine waters and lacustrine brines," Izvest. Sektora fiz.-khim. analiza (IONKh AN SSSR), 10 (1938).

E. S. Larsen and H. Berman, The Microscopic Determination of the Nonopaque Minerals [Russian translation from English] (ONTI, 1937).

E. K. Lazarenko, "Some questions on the mineralogy of sedimentary rocks," in the Collection; Questions on the Mineralogy of Sedimentary Rocks, Book 1 [in Russian] (Lvov University Press, 1954).

E. K. Lazarenko, "Questions on the nomenclature and classification of glauconite," in the Collection: Questions on the Mineralogy of Sedimentary Rocks, Books 3 and 4 [in Russian] (Lvov University Press, 1956).

A. I. Lebedeva, "Authigenic feldspars in Cambrian sands," Uch. zap. Leningr. univ., 209, seriya geol. nauk, 7, Geologiya (1956).

F. Yu. Levinson-Lessing (F. Y. Loewinson-Lessing) and E. Z. Struve, Petrographical Dictionary, ed. 2, revised and enlarged [in Russian] (Moscow-Leningrad, ONTI, 1937).

N. V. Logvinenko, "Late diagenesis (epigenesis) in the Donets Carboniferous rocks," Doklady Akad. Nauk SSSR, 106, 5 (1956).

N. V. Logvinenko and N. K. Zabolotnaya, "The identification of carbonate minerals by the staining method," Zap. Vses. min. obshch., 83, 3 (1954).

V. D. Lomtadze, "Changes in the moisture content of clays during compaction under heavy loads," Zap. Leningr. gorn. inst., 29, 2 (1953).

V. D. Lomtadze, "Stages of development in the properties of clay rocks during lithification," Doklady Akad. Nauk SSSR, 102, 4 (1955).

A. V. Makedonov, Author's abstract of dissertation: Concretions in the Vorkuta Series (Moscow, Institute of Geological Sciences, Academy of Sciences, USSR, 1954).

V. Malyshek, "Analcime and volcanic glass in sedimentary rocks and the significance of this factor in the correlation of sections of the Produktivnyi sequence on the Apsheron Peninsula," Tr. Azerb. nauchno-issled. neft. inst., 33, A collection of papers by young specialists, Baku (1936).

V. A. Makhinin, "Mineralogy of the glauconites in the Oligocene rocks of the Ukrainian crystalline shield," Min. sbornik, L'vovsk. geol. obshch., 5 (1951).

Manual on X-Ray Structural Analysis [in Russian] (Gostekhizdat, 1940).

V. I. Mikheev, "Identification of minerals in the chlorite group by x-ray methods," Zap. Vses. min. obshch., seriya II, 82, 3 (1953).

V. I. Mikheev, "X-ray determination of minerals," Zap. Vses. min. obshch., seriya II, 86, 3 (1957).

V. I. Mikheev, X-Ray Determination of Minerals [in Russian] (Gosgeolizdat, 1957).

L. M. Miropol'skii, "The origin of celestite in Permian rocks in the vicinity of Kazan and in northeastern Russia," Tr. Obshch. estestvoispyt. pri. Kazansk. univ., 51, 4 (1926).

L. M. Miropol'skii, "Mineralogical-petrographical investigation in the Zvyaginskii and Morkinskii cantons of the Mari district to study the effect of geochemical processes on ground water," Zhurnal "Marii Ilysh" (Mariiskaya zhizn' "), Ioshkar-Ola, 3 (Izd. mariisk. obl. obshch. kraevedeniya) (1930).

L. M. Miropol'skii, "Copper ores in the Permian rocks of the Tatar ASSR, and their origin," Uch. zap. Kazansk. univ., 98, Book 1, Geologiya, 10 (1938).

L. M. Miropol'skii, "Galena in Permian rocks," Izvest. Akad. Nauk SSSR, seriya geol., 3 (1940).

L. M. Miropol'skii and V. A. Polyanin, "Chalcopyrite in the Devonian rocks of southeastern Tataria," Doklady Akad. Nauk SSSR, 81, 3 (1951).

D. V. Nalivkin, "Bauxites in the Urals," Izvest. Akad. Nauk SSSR, seriya geol., 4 (1942).

S. N. Nikitin, "The Carboniferous rocks in the Moscow region and artesian waters under Moscow," Tr. Geol. kom., 5, 5 (1890).

F. A. Nikolaevskii, "Allophane minerals in the vicinity of Moscow," Izvest. Akad. Nauk, ser, 6, 6 (1912).

M. E. Noinskii, "Samarskaya Luka," Tr. Obshch. estestvoizpyt. pri Kazansk. univ. 45, 4-6 (1913).

W. Noll, Mineralbildung in System AL_2O_3-SiO_2-H_2O. N. Jahrb. Min. Abt. A, 1935, 70.

W. Noll, Synthese von Montmorilloniten.-Chem. d. Erde, 1936, 10, H. 2.

Nonmetallic Resources of the USSR, 4, Clays and Kaolin-Clays for Bleaching [in Russian] (Acad. Sci. USSR Press, 1941) pp. 20-60.

J. Orsel, Recherches sur la composition chimique des chlorites.-Bull. Soc. Franc. Min. 1927, 50, N 3-6.

L. Pauling, The structure of the chlorites. -Proc. Nation. Acad. Sci., 1930, 16.

V. P. Petrov, "Geologic-mineralogical investigations in Uralian white clays, and certain conclusions on the mineralogy and origin of clays in general," Tr. Inst. geol. nauk AN SSSR, 95, petrograf. seriya, 29 (1948).

V. P. Petrov, "Nalchikites," Uch. zap. Kabard. nauchno-issled. inst., 3 (1948).

V. P. Petrov, "The mineralogy and petrography of the non-ore raw products in the refractory and ceramic industry," in: The Physicochemical Basis of Ceramics [in Russian] (Moscow, 1956).

H. L. Piotrowskii, 1) "A new microchemical method of identifying magnesite," 2) "The identification of minerals by means of organic dyes," Sprawozdania Towarzystwa Naukowego we Lwowie, 14, 3 (1934) pp. 233-234 and 234-236.

H. L. Piotrowskii (G. L. Piotrovskii), "A new method of studying carbonate rocks by means of organic dyes," Zap. Vses. min. obshch., seriya II, 85, 2 (1956).

V. A. Polyanin and I. N. Gorizontova, "Copper ores in the Kirov district," Tr. Kirovsk. obl. nauchno-issled. inst. kraevedeniya, 12, Kirov (1939).

S. P. Popov, "Minerals in the ore beds of the Kerch and Taman peninsulas," Tr. Geol. muzeya AN, 4, 7 (1910).

S. P. Popov, "Kerchenites," Izvest. Geol. kom., 48, 10 (1929).

S. P. Popov, Mineralogy of the Crimea [in Russian] (Acad. Sci. Press, 1938).

K. A. Pospelova, A Conspectus of a General Course on Colloidal Chemistry for the Lectures of Academician P. A. Rebinder [in Russian] (Moscow University Press, 1950).

I. A. Preobrazhenskii, "Native iron in Timan," Doklady Akad. Nauk SSSR, 28, 7 (1940).

I. A. Preobrazhenskii, "Authigenic minerals and mineral development," Tr. Inst. geol. nauk AN SSSR, 40, petrograf. seriya, 13 (1941).

L. V. Pustovalov, "New data on the origin of the Lipetsk and Tula iron ores," Tr. Vses. geol.-razved. ob"ed., 202, Moscow-Leningrad (Gos. nauchno-tekhn. geol.-razved. izd.) (1932).

L. V. Pustovalov, "Geochemical facies and their significance in general and in applied geology," Problemy sov. geologii, 1 (1933).

L. V. Pustovalov, Ratofkite in the Upper Volga Region [in Russian] (Acad. Sci. USSR Press, 1937).

L. V. Pustovalov, The Petrography of Sedimentary Rocks, pts. 1 and 2 [in Russian] (Moscow, Gostoptekhizdat, 1940).

L. V. Pustovalov (L. W. Pustowaloff), "Secondary alteration in sedimentary rocks," Geol. Rundschau, 43, 2,, Stuttgart (1955).

L. V. Pustovalov (1), "Secondary alteration in sedimentary rocks and its geologic significance," Tr. Geol. inst. AN SSSR, 5 (1956).

L. V. Pustovalov (2), "Secondary feldspars in sedimentary rocks (A survey of the principal literature)," Tr. Geol. inst. AN SSSR, 5 (1956).

M. A. Rateev, "Beidellitic clays in the upper Maikopian deposits of Chernaya Ravine," Doklady Akad. Nauk SSSR, 96, 4 (1954).

M. A. Rateev and D. D. Kotel'nikov, "New discoveries of α-sepiolite in the Carboniferous rocks of the Russian platform," Doklady Akad. Nauk SSSR, 109, 1 (1956).

V. N. Razumova, "Data on the petrography of clays, pt. 1," Tr. VNIIMS, 137 (1939).

N. V. Rengarten, "Authigenic analcime in sandstones of the Kazan series in the Kirov district," Zap. Min. obshch., 69, 1 (1940).

N. V. Rengarten, "A zeolite of the mordenite group from the marine rocks of Late Cretaceous and Paleogene age on the eastern slope of the Urals," Doklady Akad. Nauk SSSR, 49, 8 (1945).

N. V. Rengarten, "Laumontite and analcime from Lower Jurassic Rocks in the northern Caucasus," Doklady Akad. Nauk SSSR, 70, 3 (1950).

J. Rodgers, Distinction between calcite and dolomite on polished surfaces. – Amer. Journ. Sci., 1940, v. 238, N 11, pp. 788-798.

E. V. Rozhkova and N. V. Solov'ev, "Experimental studies on the conditions of formation of iron-aluminum pisolitic ores," Tr. Vses. inst. min. syr'ya, 111 (1936).

N. I. Rudenko, "Exogenetic galena from the oxidized zones of sulfide deposits," Zap. Vses. min. obshch., seriya II, 83, 3 (1954).

L. B. Rukhin, Fundamentals of Lithology [in Russian] (Gostoptekhizdat, 1953).

D. G. Sapozhnikov, "Cupriferous sandstones in the western part of central Kazakhstan," Tr. Inst. geol. nauk AN SSSR, 3, geol. seriya, 28 (1948).

D. G. Sapozhnikov and I. P. Zlatogurskaya, "Cupriferous sandstones in the basin of the Ishim River in Kazakhstan," Byull. Mosk. obshch. ispyt. prirody, otd. geol., 28, 6 (1953).

I. D. Sedletskii, "The origin of colloidal minerals in soils in connection with types of soil formation and weathering," Priroda, 1 (1938).

I. D. Sedletskii (1), "The origin of montmorillonite and kaolinite, and the conditions under which they are now found in colloidal soils and clays," Doklady Akad. Nauk SSSR, 22, 8 (1939).

I. D. Sedletskii (2), "The soil absorbing complex, a paragenetic system (colloidal) of minerals," Doklady Akad. Nauk SSSR, 23, 3 (1939).

I. D. Sedletskii (3), "Gedroitzite in solonets (alkaline soil)," Doklady Akad. Nauk SSSR, 23, 6 (1939).

I. D. Sedletskii (4), X-Ray Patterns of Soil [in Russian] (Acad. Sci. USSR Press, 1939).

I. D. Sedletskii (1), "A classification of minerals in the zone of weathering," Sov. geologiya, 3 (1941).

I. D. Sedletskii (2), "The paragenesis of elements and minerals in the colloids of soils and clays," Doklady Akad. Nauk SSSR, 30, 2 (1941).

I. D. Sedletskii (3), "Paragenetic groups of minerals in the principal soil types," Doklady Akad. Nauk SSSR, 32, 6 (1941).

I. D. Sedletskii (4), X-Ray Tables for Identification of Colloidal Minerals in Soils [in Russian] (Acad. Sci. USSR Press, 1941).

I. D. Sedletskii (1), "A subdivision of the colloidal-dispersed minerals of the montmorillonite group," Doklady Akad. Nauk SSSR, 34, 4-5 (1942).

I. D. Sedletskii (2), "The composition of the colloidal-dispersed minerals in marine muds, and the problems of diagenesis," Doklady Akad. Nauk SSSR, 36, 1 (1942).

I. D. Sedletskii, Colloidal-Dispersed Minerals [in Russian] (Acad. Sci. USSR Press, 1945).

N. N. Serb-Serbina and P. A. Rebinder, "The development of structure in aqueous suspensions of bentonitic clays," Kolloidnyi zhurnal, 9, 5 (1947).

D. P. Serdyuchenko, "Two iron chlorites," Doklady Akad. Nauk SSSR, 60, 3 (1948).

D. P. Serdyuchenko, "Chlorites, their chemical constitution, and their classification," Tr. Inst. geol. nauk AN SSSR, 140. Mineralogogeochimicheskaya seriya, 14 (1953).

D. P. Serdyuchanko, Sepiolites, Palygorskites, and "Attapulgites," - Mineralogical Collection No. 9 [in Russian] (Lvov University Press, 1955).

D. P. Serdyuchenko, The Classification of the Montmorillonite Minerals—Mineralogical Collection No. 10 [in Russian] (Lvov University Press, 1956).

N. S. Shat-skii, Phosphorite-Bearing Formations and the Classification of Phosphorite Deposits [in Russian], Transactions of the Conference on Sedimentary Rocks, 2 (Moscow, Acad. Sci. USSR Press, 1955).

S. A. Shchukarev and T. A. Tolmacheva, "The colloidal-chemical theory of salt lakes," Zhurnal Rus. fiz.-khim. obshch., chast' khimich., 62, 4 (1930).

M. S. Shvetsov, Petrography of Sedimentary Rocks [in Russian] (The Mining Geology and Petroleum Press, 1934).

M. S. Shvetsov, Petrography of Sedimentary Rocks, 2d ed., revised [in Russian] (Gosgeolizdat, 1948).

A. V. Sidorenko, "Discoveries of picite in the Kerch iron ores," Doklady Akad. Nauk SSSR, 43, 6 (1944).

N. V. Solov'ev. "Investigations on the origin of dolomitic flour in the region of construction of the Kuibyshev dam," Doklady Akad. Nauk SSSR, 30, 6 (1941).

L. O. Stankevich, Mineralogy of the Sedimentary Ore Deposits in the Southern Ukraine and in the Crimea [in Russian] (Dnepropetrovsk Mining Institute, 1954).

N. M. Strakhov, "Dolomitic sediments in Lake Balkhash, and their significance in understanding the processes of dolomite formation," Sov. geologiya, 4 (1945).

N. M. Strakhov, "Iron-ore facies and their analogues in the history of the earth," Tr. Inst. geol. nauk SN SSSR, 73, geol. seriya, 22 (1947).

N. M. Strakhov, "Diagenesis of sediments and its significance in sedimentary ore formation," Izvest. Akad. Nauk SSSR, seriya geol., 5 (1953).

N. M. Strakhov, "On understanding diagenesis," in the Collection: Questions on the Mineralogy of Sedimentary Rocks, Books 3 and 4 [in Russian] (Lvov University Press, 1956).

V. A. Sulin, The Waters of Petroleum Deposits in the USSR [in Russian] (ONTI, 1935).

Tables of Interplanar Distances [in Russian] (Leningrad Mining Institute Press, 1950).

V. B. Tatarskii, "Dedolomitization and related problems," Vestnik Leningr. univ., 1 (1953).

V. B. Tatarskii, The Microscopic Determination of Carbonates in the Calcite and Aragonite Groups [in Russian] (Gostoptekhizdat, 1955).

G. I. Teodorovich, "Carboniferous rocks in the vicinity of the Dobryatino siding on the Moscow-Kazan railroad," Byull. Mosk. obshch. ispyt. prirody, otd. geol., 9, 3-4 (1931).

G. I. Teodorovich, A Geologic Sketch of the Region of the Zelenyi Dol—Ioshkar-Ola branch railroad within the Mari AO [in Russian] (Mari AO, Nos. 1 and 2, 1932).

G. I. Teodorovich (1), "Siliceous formation in the Upper Paleozoic rocks on the western slope of the Southern Urals," Byull. Mosk. obshch. ispyt. prirody, otd. geol., 13, 4 (1935).

G. I. Teodorovich (2), Descriptions of Several Thin Sections and a Brief Microscopic-Petrographic Discussion of the Phosphorites in the Blyavinskii Region [in Russian] (NIUIF Fund, No. 2067, 1935).

G. I. Teodorovich, "Petrography of the Tamdy phosphorite deposit," Min. syr'e, 10 (1936).

G. I. Teodorovich (1), "The origin of Khalilovo-type iron ores at the Novo-Troitsk deposit," Byull. Mosk. obshch. ispyt. prirody, otd. geol., 17, 2-3 (1939).

G. I. Teodorovich (2), "The Khoper beds in the Orsk-Khalilovo region," Doklady Akad. Nauk SSSR, 25, 4 (1939).

G. I. Teodorovich, "Lithology of the Tournaisian-Visean calcareous-siliceous, argillaceous sequence in the Akkermanovo-Khabarninskii region (Southern Urals)," Izvest. Akad. Nauk SSSR, seriya geol., 2 (1941).

G. I. Teodorovich (1), "Dolomitization of reef formations in the Ishimbai petroleum region." Doklady Akad. Nauk SSSR, 34, 6 (1941).

G. I. Teodorovich (2), "The mantling Jurassic deposits in the region of the Khalilovo iron-ore deposits," Tr. Inst. geol. nauk AN SSSR, 67, seriya rudn. mestorozhd., 6 (1942).

G. I. Teodorovich (1), "Minerals of sedimentary rocks as indicators of the physicochemical environment," in the Collection: Questions on Mineralogy, Geochemistry, and Petrography—dedicated to the memory of Academician A. E. Fersman [in Russian] (Acad. Sci. USSR Press, 1946).

G. I. Teodorovich (2), "The origin of dolomite in sedimentary rocks," Doklady Akad. Nauk SSSR, 53, 9 (1946).

G. I. Teodorovich, "Sedimentary geochemical facies," Byull. Mosk. obshch. ispyt. prirody, otd. geol., 22, 1 (1947).

G. I. Teodorovich, "The siderite geochemical facies in the sea, and, in general, saline waters in oil production," Doklady Akad. Nauk SSSR, 69, 2 (1949).

G. I. Teodorovich, Lithology of the Paleozoic Carbonate Rocks in the Ural-Volga Region [in Russian] (Moscow-Leningrad, Acad. Sci. USSR Press, 1950).

G. I. Teodorovich, "Basic systematic patterns in sedimentation," in the Collection: The Status of the Science of Sedimentary Rocks [in Russian] (Acad. Sci. USSR Press, 1951).

G. I. Teodorovich, "Geochemical and other conditions favorable for the formation of petroleum bitumens," Neftyanoe khozyaistvo, 12 (1952).

G. I. Teodorovich (1), "Source rocks for petroleum," Neftyanoe khozyaistvo, 8 (1954).

G. I. Teodorovich (2), "Sedimentary geochemical facies along the profile of the oxidation-reduction potential," Doklady Akad. Nauk SSSR, 96, 3 (1954).

G. I. Teodorovich, "The origin of sedimentary calcareous-dolomitic rocks," Tr. Inst. nefti AN SSSR, 5 (1955).

G. I. Teodorovich, "Sedimentary mineral-geochemical facies," in the Collection: Questions on the Mineralogy of Sedimentary Rocks, Books 3 and 4 [in Russian] (Lvov University Press, 1956).

A. I. Tsvetkov and E. P. Val'yashikhina, "Data on the thermal investigation of minerals, III, Micas," Tr. IGEM AN SSSR, 4 (1956).

W. H. Twenhofel et al., Treatise on Sedimentation [Russian translation from English] (Moscow-Leningrad, ONTI, 1936).

M. A. Usov, "Facies and phases of intrusives," Izvest. Sib. otd. Geol. kom., 4, 3 (1924).

N. A. Uspenskii, "Iron chlorites in the Alapaevsk iron-ore deposit," Tr. Lomonos. inst. geokhimii i mineralogii, AN SSSR, 7, (1936).

J. H. van't Hoff et al., Investigations on the Formation of Oceanic Salt Deposits, with Special Regard to the Stassfurt Salt Deposits [Russian translation from the German edition of 1912] (Leningrad, ONTI, 1936).

V. S. Vasil'ev, "Mordenite in the Mesozoic-Cenozoic deposits along the lower Volga and in western Kazakhstan," Doklady Akad. Nauk SSSR, 45, 1 (1945).

N. E. Vedeneeva and M. F. Vikulova, A Method of Investigating Clay Minerals by Using Dyes, and Its Application to Lithology [in Russian] (Gosgeolizdat, 1952).

N. E. Vedeneeva and M. F. Vikulova, A Method of Investigating Clay Minerals by Means of Dyes (Spectrophotometric Analysis) [in Russian] (Lvov University Press, 1956).

V. I. Vernadskii, The History of the Minerals in the Earth's Crust, 1, 1 [in Russian] (Leningrad, Scientific Chemical-Technical Press, 1925).

V. I. Vernadskii, History of the Minerals in the Earth's Crust, 2, pt. 1, 1, History of Natural Waters [in Russian] (Leningrad, ONTI, 1933; 2, 1934; 3, 1936).

M. F. Vikulova, Electron Microscopic Investigation of Clays [in Russian] (Gosgeolizdat, 1952).

I. Walter, Einleitung in die Geologie als historische Wissenschaft. I. Bionomie des Meeres; III. Lithogenesis der Gegenwart. Jena, 1893-1894.

A. N. Winchell, Optical Mineralogy [Russian translation from English] (Moscow, IL, 1949).

A. N. Winchell and H. Winchell, Optical Mineralogy [Russian translation from English] (Moscow, IL, 1953).

X-Ray Methods of Identification and the Crystalline Structure of the Clay Minerals, A Collection of Papers [Russian translation from the English] (Moscow, IL, 1955).

A. N. Zavaritskii, "Gypsum and Anhydrite from Okhlebinino," Izvest. Geol. kom., 43, 7 (1924).

A. N. Zavaritskii, "Oolitic structure," Tr. Min. muzeya AN SSSR, 3 (1929).

A. N. Zavaritskii and V. I. Mikheev, "Ktypeite," Doklady Akad Nauk SSSR, 63, 6 (1948).

P. A. Zemyatchenskii, "Feldspathization of limestones," Izvest. Akad. Nauk, seriya VI, 10 (1916).

E. F. Ziv, "Rutilization of ilmenite in a supergene environment," Izvest. Akad. Nauk SSSR, seriya geol., 12 (1956).